Bumblebee Economics

BERND HEINRICH

HARVARD UNIVERSITY PRESS • Cambridge, Massachusetts, and London, England

Library of Congress Cataloging in Publication Data

Heinrich, Bernd, 1940–
   Bumblebee economics.

   Bibliography: p.
   Includes index.
     1. Bumblebees—Ecology. 2. Bioenergetics.
3. Insects—Ecology. I. Title.
QL568.A6H39   595.7′99   78-23773
ISBN 0-674-08580-9 (cloth)
ISBN 0-674-08581-7 (paper)

To Kitty and Erica

Acknowledgments

Science is a social enterprise, not a solitary endeavor, and I owe considerable gratitude to the institutions and people that have helped me to produce what is here presented. I am particularly grateful to the United States National Science Foundation for providing me and others the material assistance without which most of the reported research could not have been done.

Part of this book took shape as a result of a trip I made in 1976 to observe the spring back home in Maine and to run in the Boston Marathon. In order to finance the trip I gave seminars describing my work with bumblebees at various universities on my way to and from the East. Following these talks there were always stimulating and fruitful discussions with graduate students and faculty. Because of Edward O. Wilson's enthusiasm for both social insects and running, I felt encouraged to spend my fall 1976 sabbatical leave from Berkeley at the Museum of Comparative Zoology at Harvard, where I studied the bumblebee's foraging behavior from the perspective of sociality and began to consolidate the research into a small book. I was honored with a Guggenheim Fellowship, and the congenial company on the fourth floor of the museum provided the final stimulus needed to get the book launched. In addition to Ed Wilson, I thank Virginia D. Adams, Tracy Allen, George A. Bartholomew, Astrid and James H. Brown, Sydney Cameron, George C. Eickwort, Adrian B. Forsyth, Bastiaan J. D. Meeuse, Chris Plowright, Peter H. Raven, and Thomas D. Seeley for comments and for critical readings of the various drafts of the manuscript. Celeste Green and Phyllis Thompson did the illustrations. Harry Foster and William Bennett of Harvard University Press provided guidance, criticism, and encouragement. Karen Bailey and Kathy Horton provided invaluable help in typing. Lynn Best, Paulette Bierzy-chudek, and David Inouye allowed me to use unpublished information from their research. To all I am grateful.

Some of the illustrations have been previously published, often in a somewhat different form, and I want to thank the following publishers and editors for permission to reproduce material for which they have copyright: American Association for the Advancement of Science (*Science*), Duke University Press (*Ecology* and *Ecological Monograms*), W. H. Freeman and Company (*Scientific American*), Macmillan Journals Ltd. (*Nature*), Sigma Xi (*American Scientist*), the Society for Experimental Biology (*Journal of Experimental Biology*), the Society for the Study of Evolution (*Evolution*), Springer Verlag (*Journal of Comparative Physiology*), the University of Texas Press (*Coevolution of Animals and Plants*), and the Kansas Entomological Society (*Journal of the Kansas Entomological Society*). Holt, Rinehart and Winston kindly provided permission to use epigraphs from Robert Frost's poems, and Alfred A. Knopf granted permission to quote Sigurd Olson.

This book will have served its purpose if biology students can gain from it a sense of how the multidisciplinary approach can be helpful to an understanding of the economic laws that are fundamental to success in living and that operate at all levels of biological organization from the molecular to the ecological. It will have been well worth the effort if it provides laymen with a view of how scientists, as well as bumblebees and their kin, do what they do, and what it means to each of them.

Contents

If I knew all there is to know about a golden
arctic poppy growing on a rocky ledge in the
Far North, I would know the whole story of
evolution and creation.

—Sigurd F. Olson, *Reflections from the
North Country*

Introduction

While growing up in central Maine in the early 1950s I spent much
time out of doors with old-timers who taught me the fascinating art of
"lining" bees. That a bee we had fed a mile from its hive could communicate to others the location of our honey-baited bee-box was an
incomparable wonder that left a deep impression. No less impressive
were the big hollow bee-trees in the forest, and the buckets full of hon-
eycomb we pulled out as the blue autumn sky was crisscrossed, like a
kaleidoscope, by thousands of humming bees.

Almost twenty years later I returned to our family farm in Maine with
a Ph.D. in zoology from UCLA and an electric thermometer, rather
than a bee-box, with which to pursue bees. I quickly put the device to
use measuring the body temperatures of bumblebees. The neighbors
were skeptical—they claimed they didn't know bees ever had a temperature! I also examined the bees' honey-crop contents. Once,
while I was pulling the abdomens off bees picked from flowers, a
stranger in overalls and suspenders demanded: *"What* in the *hell* are
you doing?" "Killing bumblebees," I replied. "Oh," he paused, "for a
minute there I thought you were one of them damn biologists; they'll
do almost *anything.*" Back in Berkeley one of my academic colleagues
asked me the same question, leaving the expletive unspoken. I told
him that I was studying foraging behavior. To this he said, "Oh—you
mean you want to find out if they go to where there is more nectar?"
Of course there is considerable truth in both assertions. But the first is
possibly a slight exaggeration, while the second is definitely a flagrant
oversimplification.

This communication gap inspired me to try to tie my research in with everyday experiences and concepts so that both the interested laymen and the interested professional would be able to share with me some of the excitement and wonder of the fascinating biology of the bumblebee as it was being revealed in field and laboratory work. I also wanted to show the natural continuity between research in physiology, behavior, and ecology. In addition, there were aesthetic reasons for writing this book. To see the patterns—to see how everything is connected to everything else—is to perceive beauty and feel harmony with the natural world. I have included numerous speculations, ideas, inferences, and personal ways of looking at the world. The opportunity to do this has been in itself a powerful spur to write.

Early in my studies it became clear that the bee's physiology is related to energy balance, and that energy balance is tied in with pollination and the reproduction of plants. Bumblebees are very often in an energy crisis of far greater magnitude than anything humans ordinarily experience. Energy economics is a major factor governing much of their biology, and this provided a theme for organizing the book.

Few aspects of bumblebee life, or human life, can escape the pervasive influence of economics. Economics can be defined as the study of the acquisition or production, distribution, and consumption of goods and services. But what makes economics so compelling and important a science is not just its breadth but its urgency: resources, or goods and services, seldom, if ever, exceed or even keep up with needs and wants. As a consequence, "economy" is almost synonymous with "frugality."

Economics has traditionally dealt only with the material welfare of mankind. This welfare is related to various possible mechanisms for the distribution of resources and services, and it is never far removed from politics. Furthermore, since economics also concerns the efficient utilization of machines for production and transportation, it is closely related to technology.

Animals, particularly those unable to regulate their rates of reproduction, are commonly forced to practice strict energy economy, but they have no politics or technology, as such. Except for those living in groups, the primary allocation of energy and resources is between different parts and activities of the same individual, and economics is a matter of efficient physiological and behavioral functions. The social insects, however, have mechanisms for allocation of resources within the colony, and thus we can see in them evolved behavior patterns that are analogous to our political systems (particularly communistic ones), in the same way that morphological and physiological adaptations that have evolved for energy economy are analogous to our technological advances.

We marvel at the social insects' sophisticated solutions to problems like our own, particularly since they seem so "rational," a quality that few humans possess but all pride themselves in. They have evolved through eons of natural selection, which operates wholly according to "reason" (utilitarian purposes). However, like the Houyhnhnms in Jonathan Swift's *Gulliver's Travels*, although the bees lead strictly "rational" lives, they cannot reason. Swift wrote, "As these noble Houyhnhnms are endowed by nature with a general disposition to all virtues, and have no conceptions or ideas of what is evil in a rational creature; so their grand maxim is to cultivate Reason, and to be wholly governed by it."

Bees evolved from their wasplike ancestors some 100 million years ago, while *Homo* has been on the scene for scarcely 2 million years. Yet, we consider these creatures to be below us on the evolutionary scale. If we reflect, however, that our evolutionary history (as apes and humans) is exceedingly brief in comparison with theirs (as bees), and that they have evolved very far along entirely different lines, then we can see that from the insect's perspective, we may well be lower. Their society includes queens, scouts, and workers, wondrous divisions of labor, and communication systems used in complex strategies of offense and defense. Various cost-benefit functions must be balanced in making decisions that are crucial to their struggle to secure, and survive on, sometimes scarce resources. Because of these analogies between their society and ours, certain anthropomorphisms crop up facilely when describing social insects. However, the great evolutionary gulf between insects and man—which makes it all the more fascinating that any comparisons at all can be made—also means the analogies must obviously be taken with some grains of salt.

My overall aim in this book is to explore biological energy costs and payoffs, using the bumblebee as a model. The physiological and behavioral bases of the energy economy of the bumblebee are traced to their wider ecological implications. I have drawn heavily on recent research results, and this book reflects the emphases and gaps in that research. The book describes the environment's effect on the bees and the bees' effect on the environment. The chapters on the colony cycle in the context of the environment (Chapter 1) and on the colony economy (Chapter 2) provide a natural backdrop for a detailed examination of physiology and behavior, and their ecological consequences, in subsequent chapters. Bumblebees are social insects living predominantly in regions of low temperature. They owe their success, in large part, to their remarkable thermoregulatory capacities. In Chapters 3–6 I discuss in detail many aspects of their thermoregulatory physiology that relate directly to their foraging activity and to their colony economy. Thermoregulation has economic costs and benefits that are balanced during foraging, as described in Chapter 7. The foraging strategy of the bees, from the standpoints of individual initiative, competition, and colony strategy, is explored in Chapters 8–10 (the word "strategy" implies purpose; however, as used in evolutionary biology it simply connotes a set of morphological, physiological, and behavioral adaptations that function to aid survival in a particular environment). Finally, the bees' exploitation of their resource environment has innumerable consequences, both from immediate and long-term evolutionary perspectives (Chapters 11–12). Their pollination of plants triggers a vast and possibly incalculable series of ecological interactions. I have interjected examples involving other organisms than bumblebees in order to illustrate specific ideas or to provide contrasts and contexts. I hope that these digressions will promote breadth of outlook, rather than causing distraction.

Although research proceeds from one step to another, the sequence taken is often not as logical nor as linear as the book that results. I had, in the beginning, no grand model or design in my mind, to be corroborated by a planned set of experiments in the field and in the laboratory. I pursued only small questions that seemed interesting in light of previously collected data. The central theme presented here—the theme of economics based on energetics—emerged of its own accord.

This book is obviously not merely about bumblebees. Bumblebees are simply aesthetically pleasing, common, and convenient animals to work with in investigating many interesting problems. Bumblebees are assuredly not the only important pollinators, any more than they are the only insects that regulate their body temperature. They are common and conspicuous animals of which many species can be identified in the field by their bright and varied colors. They adapt well to captivity, and they can be easily kept in the laboratory for experiments on physiology and behavior. They are large enough so that they can be tagged and visually identified as individuals in the field. They can be observed at close range without being disturbed, especially during foraging. And the resources upon which they subsist can be extracted from flowers by an investigator; the exact amounts that are available can be removed and measured for easy economic bookkeeping.

I have tried to minimize jargon and technical details in order to reach the more general reader. I present a limited amount of hard data in graph form, for those readers who will want to know the quantitative basis of my statements, but the book can be understood without the graphs.

CHAPTER ONE

But, for the point of wisdom,
  I would choose to
Know the mind that stirs
  between the wings of
Bees and building wasps.

—George Eliot, *The Spanish Gypsy*

The Colony Cycle

Bumblebees are associated with sunshine, with colorful and fragrant flowers of damp meadows, scenic mountaintops, and mysterious bogs —the boreal spruce-fringed bogs bordering sluggish brooks or quiet ponds. These bogs have their unique and interesting associations of living things. Much of each bog is a floating mat of vegetation held together by labyrinthine interdigitations of roots from small flowering shrubs, sedges, orchids, mosses, and pitcher plants. Sleek brook trout with bright red spots lurk under the floating edges. A pair of loons patrols the water surface. And each bog almost invariably has one olive-sided flycatcher calling loudly from the tip of a stunted spruce in springtime. The various organisms appear to act independently of each other, yet they are functionally interrelated.

The association of bogs and bumblebees is not fortuitous. Bumblebees are tundra-adapted insects, and the bogs are post–ice-age islands of tundralike vegetation with which bumblebees have probably been associated for millions of years. However, bumblebees are also found in all types of open areas, including fields, roadsides, burn areas, and mountain tops.

The bogs themselves are evolving. Each of them goes through a series of successions. During the last glaciation, about 20,000 years ago, layers of ice several thousand feet thick covered Canada, great portions of Minnesota, the Great Lakes region, New York, and New England. The ice sheets advanced southward like giant bulldozers, gouging sinkholes. When they retreated they left dams of rock and gravel.

Sedges were the first plants to invade the surface of the quiet waters behind the dams and in the sinkholes. The "sedge stage," then as now, is followed by the "iris stage," characterized by showy blue-flag iris, joe-pye weed, and bog goldenrod (Schwintzer and Williams, 1974). The cool, acid waters retard decay. Dead plant remains accumulate and form a floating mat of vegetation. The next stage of the plant succession is characterized by low shrubby plants of the family Ericaceae, principally leatherleaf, rhododendron, swamp laurel, bog rosemary, blueberry, Labrador tea, and cranberry (Fig. 1.1). After generations of these plants have left their remains to the accumulating mat, the "high brush stage" is attained, with its alders, willows, winterberry, and black chokeberry (Dansereau and Segadas-Vianna, 1952). Finally, in the "tree stage," spruces and larches appear and eventually bumblebee flowers are shaded out. At any one time, any one bog usually contains a number of these plant successions. Although at the water's edge the bog may be in the sedge-stage, it may change gradually to the iris, shrub, high-bush, and tree stages as one moves further from the water. The waves beat back the advance of the succession from the water's edge, while beavers often maintain the bog's upkeep from the land by felling invading trees and by building dams.

The boreal bogs in the Northeast are surrounded by dark and shady forests that choke out the flowers and the sunshine so important for the bees in summer. But the low vegetation in the bogs is not shaded, and the plants from one bog, or from a number of neighboring bogs, usually provide flowers all summer. Bumblebees visit a progression of flowering plants from early spring until fall (Heinrich, 1976a). The bog is a living system, largely undisturbed by man. It is one of the primary homes of bumblebees, and the bees are exquisitely attuned to it.

A day in the bog begins while the fog still lies low and heavy. A beaver glides silently on the water. Beady eyes scan the water level. A barred owl still booms from the forest.

The sun appears as a hazy globe. Spiderwebs are now glistening with dew. The swamp sparrows begin their trilling. The red-winged blackbirds come out of hiding in the low shrubs, ruffle their feathers, flash their red epaulettes, and become raucous. And, like a miniature helicopter, a small, furry, orange, black, and yellow object zooms out of the forest, lands on a rhododendron bush, and flits in apparent great haste from flower to flower.

It is a queen bumblebee. She has spent the last eight months,

Fig. 1.1   Photograph of Huckleberry Stream, a study area near my residence in Maine. The quiet stream is surrounded by shrubs visited and pollinated by bumblebees. At the time of this photograph—early May—no leaves have yet appeared, but the leatherleaf in the foreground is blooming and has already been visited by numerous bumblebee queens.

throughout the recent winter, hibernating underground in an almost lifeless state. She is alone. All the drones and workers from her hive, as well as her mother—the old queen—died in the fall before the snow fell. In the spring, after the snowdrifts had melted, the sun's warmth signaled her to awaken from torpor and emerge. The depth of her burrow had a marked effect upon the time when she was warmed and resumed activity (Szabo and Pengelly, 1973). Although the bee is now completely alone, the colony she came from may have produced up to a hundred virgin queens and males the previous fall. (Other species of bumblebees in areas of long growing seasons may send out as many as a thousand new queens.) Many did not survive the winter. Many more will die. Only one of the nest's new queens will, on the average, be successful in producing a colony that will yield the ultimate objective of all colonies: new queens and males.

Having fed from the rhododendron blossoms, our queen bee is flying close over the forest floor near the bog's edge. She is powered by the energy she has received from the sugar of the rhododendron's nectar. She lands on the ground at frequent intervals, crawls under leaves and into holes in the ground, and then resumes her wandering flight. She is searching for a suitable site to found a colony. Meanwhile, the protein from the pollen she has eaten is being converted to eggs in her ovaries. The bee is fertile, having been inseminated the previous fall.

After searching all day, day after day, for possibly two weeks or longer, she may find, in the damp tunnels beneath a decaying tree stump, an abandoned nest of a white-footed deer mouse, a red-backed vole, or a chipmunk. She is not choosy, however. Any dark cavity filled with fine plant fiber will do. She tugs and pulls at the grass and bark fibers, creating a cavity about 2 cm high and 3 cm wide. Some species, such as *Bombus fervidus,* may make their own nests in dense grass on the surface of the ground. After their colony gets established, bees of this species keep adding dead grass until a bulky nest is created that is indistinguishable from a field-mouse nest. In the initial nest, near the entrance to her small cavity, each bumblebee queen fashions a thimble-sized honeypot out of wax scales exuded from glands between the armored segmental plates on both the top and bottom of her abdomen. (Honeybees have wax glands only between the segments of the ventral surface of the abdomen.)

After returning to the nest from a foraging trip, our queen contracts her abdomen in a small series of jerks and regurgitates nectar from her honeycrop into the honeypot. She rubs her hind legs together, releasing the pollen loads from them and forming a pollen clump directly on the floor of the nest cavity. A batch of 8–10 eggs is laid into the pollen, and the eggs of this "brood clump" hatch into larvae that superficially resemble maggots. The larvae will ultimately spin cocoons of silk and pupate inside them.

During the next month, after having laid the first batch of eggs, the bumblebee queen spends a lot of time in the nest perching on this brood clump, which is periodically provisioned with nectar-moistened pollen. She makes frequent trips far afield to forage. After the first workers emerge, they help her in the nest, caring for subsequent broods. Eventually they take over all the foraging duties (Fig. 1.2).

As the larvae grow the brood clump expands. A number of eggs (the exact number depending on the species) are clustered in wax packets

Fig. 1.2 The initial nest of a bumblebee (*B. vagans*), built inside a mouse nest, showing the honeypot and brood clump with pupae and new egg clumps attached on top of the pupae. The first emerged worker is sipping from the honeypot. The queen is absent, presumably foraging. The original cover of the nest has been removed.

onto the outside of the first and subsequent batches of cocoons. The eggs, larvae, and pupae constitute the brood, and this is at first separate from the honey store in the honeypot. But after the first workers have chewed their way out of the cocoons, these silken cradles are then cleaned and become storage-pots for honey or pollen. In some species, the so-called "pocket-makers," however, the pollen is put into separate pockets of wax directly below the larvae. "Pollen storers" pack the pollen into the empty cocoons, to be later retrieved and placed into the brood cells with the larvae. The egg-to-adult transformation takes 16–25 days.

The larvae derive all of the proteins, fats, vitamins, and minerals necessary for their growth from pollen. The queen bee also feeds on pollen to obtain the protein she needs to produce eggs. Workers cease growth after they emerge from their cocoons, but they need pollen for a few days more. Subsequently they need only an energy supply, and they can subsist almost exclusively on sugar (derived from nectar) for the two weeks or so that constitutes their normal life span.

In a honeybee colony, each larva matures inside a separate, identical, hexagonal cell, one size "mold" being used for workers and a slightly larger one for drones. Bumblebees, by contrast, as already mentioned, usually lay their eggs in a cluster in a distensible cell. As each group of communally fed larvae grow, the cell increases in size with them. As the larvae become mature they separate and are fed individually before each spins a silken cocoon (Fig. 1.3). Some emerging workers may be tiny—less than 0.05 g, or about half the size of a honeybee—while others may weigh 0.6 g or more, almost as much as the queen. In wild colonies of many species, the workers from the first brood are smaller than those from subsequent broods (Knee and Medler, 1965). The evolutionary significance of these size differences is unknown, but based on experiments by Chris Plowright of the University of Toronto, the immediate cause is probably nutritional. Plowright removed female larvae from bumblebee nests and placed them into artificial cells—holes drilled into blocks of beeswax. He hand-reared the larvae in an incubator (33°C) by feeding them a mixture of honey, water, and pollen. He found that larvae fed to repletion at one or two hour intervals, day and night, achieved average weight gains of about four times their body weight in 24 hours. These larvae grew to become queen-sized individuals. Indeed, they could have taken on the role of queen themselves, should the old queen have been killed. But they did not mate and any offspring of their unfertilized eggs would have become males (drones). The fertilized eggs of the queen can become either workers or queens.

Queen production can be viewed as a prolongation of the normal pattern of worker growth. If feeding is stopped before larvae from fertilized eggs have attained queen size, they become workers. After being deprived of food, provided they have achieved a minimum size, the larvae pupate. Bees of any desired size, within the normally observed size range of the species, can be produced simply by food deprivation (Plowright and Jay, 1977).

In part, the bee larvae shut off their own food supply. After molting for the last time (they molt four times), larvae spin silk when they are not engaged in feeding. Thus, if they are not fed at frequent intervals they spin a restraining belt of silk about themselves, which cuts them

Fig. 1.3 Bumblebee eggs, larvae, and pupae. An egg clump has been torn open for the photograph, and the cocoon is slit to reveal the pupae, a nearly fully-formed young bee that is still unpigmented except for the eyes. The prominent tongue is as yet unfolded. The round mounds are the tops of cocoons. (Photograph by E. S. Ross.)

off from further food. They essentially close up the aperture through which they are fed. Those larvae destined to become queens are fed at frequent intervals and have no time to spin silk, and thus they can continue to feed. The application of juvenile hormone to bumblebee, honeybee, and stingless-bee larvae will also induce queenlike characteristics in the emerging female adults. High food consumption—especially the consumption of certain nutrients—may be related to hormone production in bumblebees, as it is known to be in honeybees (Röseler and Röseler, 1974).

While the first workers are still in their cocoons, the queen attaches several additional egg clusters to the sides of the cocoons. As more and more cocoons are formed, the queen deposits more and more egg clusters (Fig. 1.4). As a result, food demand and supply are balanced, in that the number of larvae that will need to be fed remains roughly proportional to the number of workers that will collect food. As the

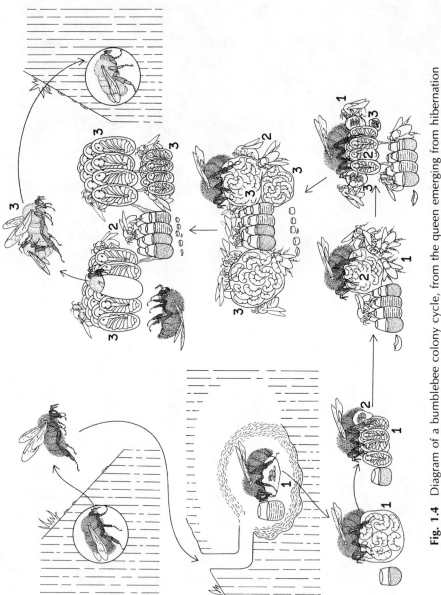

Fig. 1.4 Diagram of a bumblebee colony cycle, from the queen emerging from hibernation (left) to new queens (lightly stippled) emerging from cocoons of the third brood (eggs at lower right), mating, and hibernating (right). Note progression of eggs of specific brood packets to become larvae, pupae, and adults, and the use of the empty cocoons for honey or pollen (stippled) storage. The diagram indicates the production of two worker broods and one queen brood, the latter from three separate egg batches.

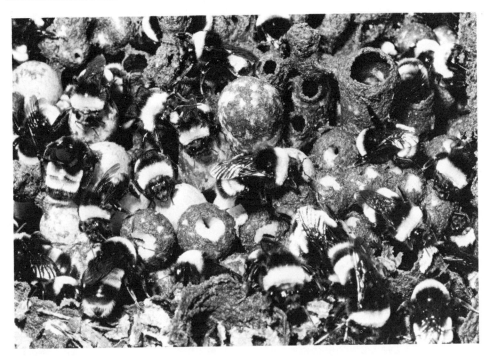

Fig. 1.5 A thriving large bumblebee colony (*B. occidentalis*) showing many cocoons and larvae and some pollen and honeypots. (Photograph by E. S. Ross.)

season progresses and the colony grows, the queen lays more eggs at a time, and lays them more frequently, until she may lay every day. In a rapidly growing colony, new workers emerge every day. The silvery-grey workers dry within a few hours after they leave their cocoons, and the bright colors of their velvety fur emerge. Two days after emerging they may leave the nest to go foraging. The edges of old cocoons are extended with wax collars and reutilized as honeypots; in some species they are also used as pollen-storage pots (Fig. 1.5).

The life expectancy of the bumblebee colony is correlated with the length of the season. Colonies of *B. atratus* of Brazil may last two or more years, with two generations of queens and drones per year. Colonies of many species in arctic and temperate regions last less than two months. In temperate regions the colony cycle terminates in late summer or fall, and there is always only one cycle of sexual production per year. The end of the colony is imminent when all of the larvae develop into new queens and males rather than workers. Males leave the colony soon after emerging. They fend for themselves and do not constitute a further drain on the colony's resources. The workers soon die off. The new queens of some species remain for a while, helping in the nest and sometimes foraging, before being inseminated and dispersing to seek underground quarters for hibernation. The old queen dies, along with the workers and drones, before winter.

We still do not know how and why the colony switches from the production of workers to that of reproductives, thus curtailing further colony growth. In Maine, I have often observed bumblebee drones concentrated in one small area, while at the same time a few miles away only workers of the same species could be seen. Variations in food availability in the two areas at critical times may have affected the colony cycles. Among many bees, the queen can produce either drone or female eggs at will by regulating the flow of stored sperm from her sperm receptacle as an egg is being laid. But how does the bumblebee queen decide to lay drone eggs, and how do the workers "decide" to rear queens rather than workers from the fertilized eggs? At present we do not know the complete answer. It appears, however, that food availability and hormones have an important influence.

As the bumblebee colony gets large, conflicts arise between queen and workers. The workers often attempt to eat her freshly laid eggs, and she has to guard them from being eaten. P. F. Röseler in Germany believes that large colony size became possible for bees through the evolution of chemical messengers (pheromones) that could be used to suppress social stress in the colony by "tranquilizing" the workers. Bumblebees exhibit more aggressive tendencies toward each other in the nest than other bees, particularly when the colonies become populous and the queen can less readily maintain her dominance.

Ultimately, colony size and the rate of colony build-up also depend on worker mortality. In Maine during August, up to 20 percent of the workers in the field are sometimes parasitized by conopid fly larvae that consume the contents of the bee's abdomen. Half of the colony's workers may die in the field every week, and many wild colonies never produce reproductives. In captive colonies, where bees are protected from parasitism, workers may live for several months, and huge numbers of reproductives can sometimes be produced.

Bumblebees do not lay up large surpluses of nectar and pollen, even though they forage for much longer hours than honeybees, and even though they visit two to three times as many flowers per unit time. Unlike honeybees and many other bees, they often forage from before daylight till after dusk, at low air temperatures as well as at high. They generally live from hand to mouth, immediately converting their food surplus into babies, although they may put aside considerable stores of pollen and honey at the point when the colony is about to initiate the production of drones and new queens. Some species, like B. affinis, B. terricola, and B. impatiens, lay up moderate stores of honey and pollen that can tide them over a few days of rainy weather. Being able to forage on most days and in a wide range of weather conditions, bumblebees have a steady economic income and have no need to save for the future, particularly when the accumulated profits would invite potential robbers like skunks and foxes. Further, unlike honeybees, they need not lay up stores to tide them through the winter.

Bumblebees are widely distributed throughout Europe, Asia, and from the Arctic Circle, 880 km from the North Pole, to Tierra del Fuego, the southernmost tip of South America. They occur in Africa north of the Sahara, and they have been introduced as pollinators of clover into Australia, New Zealand, the Philippines, and South Africa. They are scarce in deserts and hot climates, where solitary bees may be abundant, but they are often very numerous in cool temperate regions and on the summits of mountains in tropical areas. There are possibly four-hundred species worldwide, and fifty of these occur in the United States. In contrast, there are probably close to 20,000 species of other kinds of bees in the world and nearly 4,000 species in North America alone. Despite their low species number, bumblebees are often very common. A half-dozen can sometimes be caught in a sweep of the net at a favorite flower. Otto Plath, in his book on the North American bumblebees, reported finding as many as eight colonies in ten square yards of unplowed ground with thick grass. Undoubtedly, such high nest densities are highly unusual. Bumblebee nests are usually sparsely distributed and often very difficult to find, although skunks appear to have little trouble in locating, and robbing, them.

Bumblebees have some alarm and defensive behaviors not observed in either honeybees or stingless bees. When mildly alarmed, a bee that is in the nest or is perched and not flight-ready will raise her middle legs. If stimulated further, she will flip on her back and extend her legs sideways, as if to brace her body, while pointing her sting-tipped abdomen up in the air and opening her mandibles. The bumblebee's sting, unlike the honeybee's, is not barbed, and she can sting repeatedly without sacrificing her life. Bumblebees also on occasion spray feces, which could serve a defensive function. One species, *B. fervidus,* incapacitates arthropod nest invaders by covering them with regurgitated honey.

The bumblebee's sting is probably a deterrent for most predators (although some—skunks and shrikes—are known to relish bumblebees), and foraging bumblebees have not, apparently, had to evolve many mechanisms for avoiding predators. Lincoln Brower from Amherst College has convincingly shown that a toad readily snaps up a bumblebee —but not a second one! Toads learn to avoid bumblebees for life after one trial. In addition, the toad that has made one contact with a bumblebee with its stinger intact subsequently avoids all furry flies with color patterns similar to the bee's. Birds undoubtedly learn as quickly as toads, and, through millions of years of natural selection, some palatable flies, beetles, and moths have come to look like bumblebees and thus take advantage of the bees' protection from predators. Indeed, we find striking color convergence among flies of several families (Syrphidae, Tabanidae, Asilidae, and Oestridae—see Figs. 1.6, 1.7) and bumblebees (Gabritschevsky, 1926). In addition to this Batesian mimicry (where edible species mimic noxious ones), the bumblebees from some geographical regions mimic each other in a form of Müllerian mimicry (where several noxious species mimic each other), reinforcing the advertising of the common danger. Potential predators then have to learn to avoid only one color pattern rather than many.

Bombus queens just initiating their colonies probably have no greater enemies than other *Bombus* queens, possibly because of competition for nest sites. It often happens that one queen will find a desirable nest site already occupied by another queen. A fight usually ensues. Up to eight dead queens have been found at a single nest entrance. The original resident queen is sometimes killed, particularly if the first brood of workers has not yet emerged to aid in the nest defense. An invading queen, if successful in killing the nest owner, immediately accepts the brood of the displaced queen if she is of the same species, and sometimes even if she is of another species. As a result of this social parasitism, one sometimes observes colonies containing workers of two species. The supplanting species is always one that emerges later from hibernation than the species that initially estab-

Fig. 1.6 A bumblebee, *Bombus edwardsii,* sucking nectar from a composite flower. Note the long tongue and antennae. This bee is a male, or drone. Note the hairy part of the hind leg. In workers and queens the leg is smooth in the center, being adapted as a pollen-carrying apparatus, the corbiculum. (Photograph by E. S. Ross.)

lished the nest. For example, *B. affinis* sometimes supplants the earlier-nesting *B. terricola,* and *B. lucorum* sometimes supplants *B. terrestris.* In the high Arctic at Lake Hazen there are only two *Bombus* species, and one is an obligate social parasite of the other (Richards, 1973). Mixed-species colonies appear to get along peaceably, possibly because all members acquire the same colony odor. One can experimentally produce mixed colonies (of some species) simply by introducing brood from one into the nest of the other. Furthermore, adult workers (particularly those that have freshly emerged) of one species can be introduced into other nests of the same species. There may be initial fighting, but some of the strangers are usually adopted into the worker force.

Fig. 1.7 A syrphid fly, *Criorhina* sp., that mimics bumblebees. Compare the antennae and tongue with the model (Fig. 1.6). (Photograph by E. S. Ross.)

One genus of bumblebees (*Psithyrus*) specializes in taking over *Bombus* nests. *Psithyrus* comprises a small number of species that are social parasites, or "cuckoos," of *Bombus*. These bees have no worker caste, and they lack the specialized pollen-carrying apparatus (corbiculae) on the hind legs. The *Psithyrus* females enter established *Bombus* colonies, kill the resident queen, lay their own eggs, and let the *Bombus* workers rear the eggs to adulthood. The workers of one of the surface-nesting species, *Bombus fervidus,* have a defense, as already mentioned, against such parasites—they regurgitate honey onto invading *Psithyrus* females. *Psithyrus* females are heavily armored, and they are not generally stung to death unless they attempt to invade a colony that already has a large worker force. However, they gain little advantage in invading a colony with few workers, since the number of adult parasites the hosts can produce is severely limited. Thus, they have to balance safety against potential reproductive success in deciding when to invade a colony.

During the colony cycle, which I have outlined, the willows, leatherleaves, rhododendrons, blueberries, cranberries, northern winterberries, black chokeberries, roses, field spireas, and other plants flower in sequence in an orderly progression from spring to fall. Each is pollinated largely by bees, particularly bumblebees, and sets fruit (Reader, 1977). The fruit produced by the blueberry bushes in the bog is generally picked, as soon as it ripens in the fall, by robins, thrushes, and waxwings. The winterberry, rose, and chokeberry fruits remain on the twigs, sticking up out of the snow in winter. Partridge (ruffed grouse) and late migrant songbirds feed on them. The cranberries are the last to ripen. They sweeten after remaining under a blanket of snow and provide nourishment to birds the following spring. The birds, in turn, carry the undigested seeds, spreading them throughout their travels. In this way the sessile plants are able to occupy new territory and spread to niches as they become available. In the bog, the bees, birds, and plants are all functionally interrelated.

CHAPTER TWO

The prescient female rears her tender brood
In strict proportion to the hoarded food.
—Evans

Economy of the Colony

Edward O. Wilson has metaphorically and succinctly summarized the basic situation faced by a social-insect colony: "It is helpful to think of a colony of social insects as operating somewhat like a factory constructed like a fortress. Entrenched in the nest site, and harassed by enemies and capricious changes in the physical environment, the colony must send foragers out to gather food while converting the secured food inside the nest into virgin queens and males as rapidly and as efficiently as possible."

In the factory that is a social-insect colony the ultimate product is new queens and males that will go on to produce other factories. As in any factory, basic operations can be broken down into various steps, and the energy economy can be diagrammed to locate the channels of material and energy flow that lead from the raw materials in the field to the final product. Details of these various steps will be discussed in the chapters that follow. I will first indicate, in very broad strokes, the overall pattern.

Resource acquisition is accomplished solely by the workers. Workers expend energy in foraging, but they bring back more calories than they use up. In addition, they bring back pollen, the protein building material needed to make new bees. Both nectar (sugar) and pollen are deposited in communal pots. Sugar is the energy source that drives the whole system. Some of the sugar is synthesized into wax, mixed with pollen, and used as a construction material. The sugar from the honey-pots is distributed among house bees, who feed some of it directly to the larvae. The sugar's energy content is also passed indirectly to the larvae through warmth from the workers' bodies. As long as there are sufficient resources coming in, and these resources are used efficiently, the colony expands until it reaches a critical size. Its resources and invested capital are then expended in queen and drone production (Fig. 2.1).

A factory operation can be streamlined through specialization and smooth coordination of individual parts. Similarly, in the bumblebee colony there is division of labor and interdependence among the queen and her workers. Different workers operate on different steps in the production of the same product. However, whether there is "cooperation" is debatable.

It is commonly supposed that the beehive represents a collection of individuals working harmoniously for the common good. The workers are infertile daughters of one female, the "queen," and they appear to be "altruistic" in aiding her reproductive output. But in this system the individual workers are only functional appendages of the queen. They may be "altruists," but they have no choice: they can only work for the queen; they cannot produce their own offspring. If the queen were to die the workers could produce male eggs, but while she is alive her aggressiveness physiologically castrates them so that they are unable to lay eggs. If, when the colony gets large and royal control wanes, a few workers do manage to lay eggs, the queen eats them. If the queen is removed, the workers fight among themselves; the ovaries of a dominant (generally larger) worker develop, and she may then initiate egg-laying. But since workers do not mate, they can only lay unfertilized eggs, which contain half the chromosome complement (haploid), and which, in Hymenoptera, develop into males.

We generally assume that behavioral and other attributes have evolved because they in some way promote survival and reproductive success. It is not, therefore, obvious why workers should allow themselves to be manipulated into helping their mother—the queen—at the expense of their own reproductive output. How could such behavior—which entails *not* passing one's own genes on to the next generation—persist from generation to generation? One plausible explanation—a genetic predisposition that could ease resistance, provided the advantages of sociality are great enough—is that siblings share many genes with each other. Thus, one's genes can be passed on to the next generation by one's siblings as well as one's own offspring. The ge-

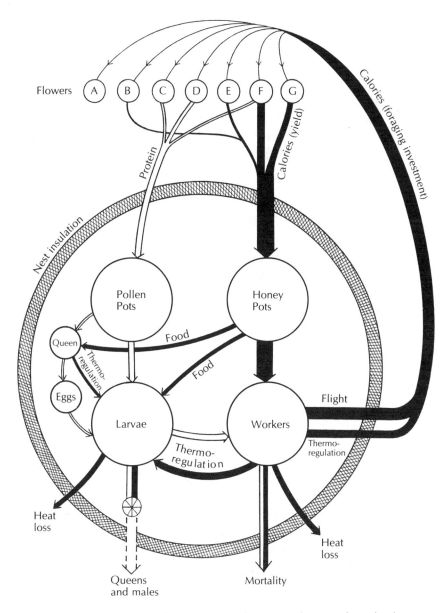

Fig. 2.1 Flow diagram of the movement of matter and energy through a bumblebee colony.

netic predisposition to aid sisters should be particularly strong in the Hymenoptera—bees, wasps, ants, and their kin. Because hymenopteran males are haploid (have one rather than two sets of chromosomes), sisters share, on the average, three-quarters of their genes (half the genes they get from their mother are the same, and *all* the genes they get from their father, since he has only one set), whereas mothers share on the average only one-half with their own offspring. Females therefore could pass more genes on to future generations by investing in care of their younger sisters than by investing in the production of offspring of their own. (In Hymenoptera, the males leave the nest and do not take part in the nest economy.) Even if this is not sufficient cause for sociality, it should still, given other selective pressures such as those arising from parasite and predator defense and food economy, at least reduce the evolutionary resistance to it.

Not all bees use the slave labor of their own offspring. Female solitary bees must each perform a wide variety of tasks, much as a bumblebee queen does in starting a colony. The females of some solitary species dig tunnels in the earth and construct a nest at the bottom. Others, such as carpenter bees (*Xylocopa* spp.)—which superficially resemble bumblebees—build their nests in holes bored into wood (Fig. 2.2). The mason bee (*Hoplitis anthocopoides*) fashions a nest out of soil, saliva, and small pebbles, building it onto the side of a boulder. Still other solitary bees hollow out plant stems (*Ceratina* spp.) or make nests in prehollowed stems (*Hylaeus* spp.). Nests of solitary bees are often lined with glandular secretions. Leaf-cutter bees (*Megachile* spp.) cut pieces of leaves or flower petals and use them for a nest lining. Each solitary bee must not only build the nest but also collect both pollen and nectar from flowers, provision the nest-cell with a mixture of both, and finally close the nest. The entombed young develop without further care. They usually do not emerge till the following year, but some species have several broods in a single year.

In social bees there is division of labor resulting in considerable specialization in, as well as out of, the nest. Some individuals work in the nest, cleaning and building cells, caring for the young, and regulating nest temperature. Others guard. Some bees specialize in collecting nectar. Others collect pollen. Most collect both nectar and pollen, but different individuals tend to specialize on different kinds of flowers. In honeybees, workers are morphologically nearly identical—they all come out of essentially identical molds; the hexagonal brood cells for

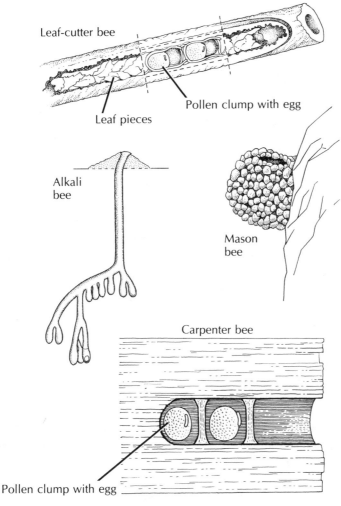

Leaf-cutter bee

Pollen clump with egg

Leaf pieces

Alkali bee

Mason bee

Carpenter bee

Pollen clump with egg

Fig. 2.2 Some of the different modes of nest construction used by solitary bees. The nest of the mason bee, *Hoplitis anthocopoides,* is constructed of pebbles glued together by glandular secretions. A leaf-cutter bee, *Megachile* sp., makes its nest in a hollow stem lined with fresh leaf pieces that envelop and separate the pollen balls of different cells. The nest of an alkali bee, *Nomia melanderi,* consists of cells branching from tunnels dug into the soil. The carpenter bee, *Xylocopa* sp., deposits its pollen and eggs into holes bored into wood.

workers, always numbering 4.83 per square inch, are also used for honey and pollen storage. Division of labor in honeybees depends largely on age (temporal polyethism). Individual workers proceed from birth to death through a succession of tasks: "house duties," nest guarding, and finally, after two weeks, foraging in the field. In contrast, in many ants and termites, division of labor among workers is based on morphological differences, such as large body size and strong mandibles, that distinguish soldiers from minor workers, for example.

A bumblebee colony is started by a single queen (as is a sweat bee colony), and she must be able to perform all of the nest duties—building comb, feeding larvae, adding onto the nest, carrying out debris, defending the nest and regulating its temperature—as well as foraging for both pollen and nectar from many different kinds of flowers, in addition to her primary function of egg-laying. Unlike most highly social Hymenoptera, which reproduce by swarming, in the primitively social bumblebees and sweat bees there is no fundamental difference in tasks between queens and workers. Qualitatively, the queen and workers do similar tasks, although as the colony cycle unfolds, the queen is more and more restricted to building brood cells and ovipositing in them (Sakagami, 1976).

In bumblebees there is no strong age-dependent division of labor among the workers. There are only two broad divisions, foraging and hive duties, but tasks are readily interchangeable. Shôichi F. Sakagami and Ronaldo Zucchi (1965), who kept marked individuals of the Brazilian *B. atratus* under observation for up to two months in a captive colony, observed that there were great individual differences. But any one worker may perform all of the different tasks in a single day, although she is likely to concentrate on hive duties up to 10–15 days of age, and on foraging thereafter.

Foraging and hive duties are, in large part, allocated on the basis of body size (Brian, 1952), which is determined in the larval stage by nutrition. Some larvae, pushed to the periphery of their communal batches, receive less food than others more favorably placed. Underfed larvae become small workers (Plowright and Jay, 1977). The smallest bumblebees of a colony, often no larger than large flies, may never forage at all. The large size-range among the workers of a colony promotes division of labor, since the small bees can walk through the intricate galleries of the nest and be useful in the hive, while the large workers, the primary foragers, are better able to regulate their body

Fig. 2.3 A bumblebee worker (*B. occidentalis*) examining a pollen pot before dropping her pollen loads into it. (Photograph by E. S. Ross.)

temperatures in the field and to fly in strong winds. On the high mountains on the Paramo of Costa Rica, for example, *B. ephippiatus* queens may forage all day despite cold and winds, while small workers come out to forage only near midday when it is warm (O. R. Taylor, personal communication).

The larger bees can also visit more flowers with deeper corolla (and ample nectar supply) than smaller bees (the "division of labor" among colony-mates in the kinds of flowers they utilize in the field is discussed in detail in Chapter 9).

In bumblebees, all individuals of a nest appear to work independently. Unlike other hymenoptera that are more highly social, bumblebees never exchange food. However, bumblebees are often very fastidious as to where in the nest they deposit their pollen and nectar. They examine numerous honey and pollen pots before unloading (Fig. 2.3), and this exploratory behavior may play a role in the assessment of colony needs. In honeybee colonies, by contrast, foragers regurgitate their nectar to receiver bees, who accept it or reject it depending on what the colony needs, and who thus indirectly communicate hive demand and need to the foragers.

Once committed to a task, the bees persist at it. Young bees are most responsive to the colony's changing needs. I found that in a colony with ample stores of pollen and honey, all of the workers became nectar foragers within three to four days after emerging from their cocoons. However, later on, when these bees had accumulated food stores in the nest, the majority of newly emerged workers remained in the nest and did not forage. Some of them made orientation flights, but they spent much time loitering at the entrance in the typical stance of guard bees. Meanwhile, the bees that had previously become foragers never hesitated an instant at the nest entrance. They continued to forage every day on successive trips. New workers emerge at frequent intervals, and these bees are likely to perform whatever tasks are still open to them.

In both honeybees and bumblebees, the division of labor is highly flexible. Bees change their tasks in accordance with what needs to be done. While making some routine observations on a colony of *B. vosnesenskii* in Berkeley, we were surprised to find that even the new queens will, under some circumstances, assume the major foraging duties in their parent colonies. When we originally found and observed the colony in early June, it contained the old queen, 260 workers, 140 new queens, and 1,020 eggs, larvae, and pupae. At that time, new queens returning to the colony always had an empty honey crop and carried no pollen. However, in the declining colony, when only 15 workers and 220 eggs, larvae, and pupae remained, the colony's 26 new queens brought in most of the colony's daily supply of pollen and honey.

How does colony size affect operating capacity? There are different costs and payoffs. The larger the colony, the less each individual has to contribute to defense, nest-temperature regulation, and other aspects of nest maintenance, but the more the resources are potentially limiting. In colonies that are large, normal activity usually produces enough heat to warm the nest. In the early stages of colony founding, or in small nests, on the other hand, some individuals have to take time out to "incubate" (see Chapter 5 for details).

The "fortress" aspect of the insect "fortress factory" is also enhanced by large group size. This can intuitively be verified by anyone trying to keep track of two (rather than one) bumblebees attempting to sting. The food stored for times of scarcity, as well as the highly palatable young, offer tempting targets for predators and parasites, and social insects have evolved formidable defenses. Charles D. Michener, a foremost authority on bees, believes one of the prime movers for the evolution of sociality in bees has been the need for defense against predators and nest parasites. In bees, all workers are potentially able to defend the nest, although some individuals, as already mentioned, are behaviorally more specialized as "guards" than others.

There is, however, a potent limitation on sociality and large colony size. All organisms that form societies, whether they be bees, ants, or humans, face increasing challenges in the procurement of raw materials as group size increases. This is because the physical requirements of the group increase linearly with increasing numbers, while the energy yield available from the home area to each individual decreases with group size. Colony size or energy use by the colony can increase only by utilizing a larger home area, increasing the energy yield from the home area, or increasing the efficiency of conversion of the energy resources gathered (Hamilton and Watt, 1970). Thus, there is strong selective pressure on social organisms to perfect their energy procurement, processing, and conservation mechanisms.

In bumblebees, maximum colony size depends in part on the length of the growing season. Arctic bumblebees have adapted to a short, cool, flowering season by producing large first broods of workers (about sixteen) (Milliron and Oliver, 1966). The bees forage more or less continuously throughout the day and into the arctic night to satisfy their food and energy demands. Most bumblebee colonies in temperate regions produce about eight workers in their first brood, and the nests ultimately contain from about 50 to 400 bees at any one time, although captive colonies containing up to 1,600 bees have been reared (Horber, 1961). The largest wild bumblebee colony located in North America was one of *B. impatiens* in Michigan. On August 26, 1975, it contained 756 active bees and 385 larvae and pupae (Husband, 1977). The few bumblebee species that live in Central and South America, where growing seasons are long and the colony cycle may last up to two years, can attain colonies of up to 2,000 bees.

Social insects have evolved ingenious ways to maintain continuous energy balance despite huge colony size. Food storage, practiced by honeybees, stingless bees, and some ants, is one method. Another is that of the termites, which use symbiotic protozoa and bacteria that help them extract energy from plentifully available wood and other cellulose fiber.

The social bees rely on food resources that are scattered very widely in tiny packets. Thinly spread resources can be depleted rapidly, and bees are tied to a nest, although honeybees, particularly the African variety, are known to leave en masse to set up a nest at a new location if food becomes depleted locally. The bees' primary advantages in harvesting large amounts of nectar and pollen in short time are rapid flight and efficient communication. The latter allows rapid and massive hive response to changing resource availability.

The bumblebees' colony cycles are under severe time constraints: the virgin queens and males must be produced at the end of one season. Energy input to the colony must be at a high and continuous rate to produce many workers rapidly. The more workers a colony can produce throughout the season, the more queens and males—the determinants of colony fitness—will be produced at the end of the colony cycle.

Any adaptation that prolongs the bees' ability to forage should be advantageous. One of these adaptations involves their eyes. Social bees, such as bumblebees, are able to leave the hive early in the morning and to return late in the evening, when landmarks are no longer visible, because they are able to use the polarized light of the sky for homing. Their three ocelli (small, "simple" eyes), located on top of the head between the two large main pair of eyes, are of particular importance in this orientation (Wellington, 1974). Without the use of these eyes, bees begin to forage later and cease sooner than normal workers. With ocelli intact, workers can spiral to get their bearings from the sky and make a beeline for the nest even when landmarks are no longer visible. Without the use of the sky's polarized light, they take longer to get back; they zig-zag, using landmarks for orientation, or, lacking visible landmarks, as in the late evening, they may be stranded at the foraging site.

Like other social bees, the bumblebees rely on the tiny food droplets scattered in widely dispersed flowers. However, unlike honeybees and stingless bees, the bumblebees do not communicate the distance and location of potential food sources to colony mates. Bumblebees—which live primarily in arctic and temperate regions—excel in one important aspect of physiology and behavior. They have evolved superb mechanisms of thermoregulation (Chapters 4 through 7) that allow them to fly, forage, and care for their brood under harsh weather conditions, where all other bees are excluded. In the Arctic and on mountain tops, for example, air temperatures may remain continuously below 10°C, even in summer, and bumblebees have been seen to fly at −3.6°C, even in a snowstorm, and in wind and rain (Bruggemann, 1958). Most solitary bees and honeybees are unable to forage at air temperatures less than 16°C. If low temperatures persist, colony growth must stop.

Relatively little is known about how the various social insects manage their energy resources. Bumblebees appear to require a great deal of energy to keep their colonies growing. But the actual amounts and their allocations are not known, since the bees generally live from hand to mouth. A deeper understanding of the hive economics will undoubtedly come from patient observations of the labor of individuals in a variety of internal and external circumstances.

Some of the ideas on the hive economy of bumblebees came into sharper focus during a set of observations made by Tracy Allen, Sydney Cameron, Ron McGinley, and myself (1978) on a large colony of *B. vosnesenskii* near Berkeley. We made an analysis of the rate of food input to the colony, comparing it to the food reserves and the worker force, in order to arrive at a measure of energy flow into and through various components of the colony.

We observed the traffic in and out of the colony for one whole day from 5:00 A.M. until 9:00 P.M. During that day, we observed 1,932 bee foraging trips, and 958 of the returning workers carried pollen (Fig. 2.4). On the next day, we sacrificed samples of bees entering and leaving the nest, weighing the pollen loads and dissecting out the honeycrops to measure volumes and concentrations of syrup carried out of and into the colony. The mean amount of sugar carried by workers was 0.0021 g on leaving and 0.027 g on returning. The mean pollen load per worker was 0.021 g. On the basis of the total foraging trips per day, we calculated that the net daily sugar intake to the colony amounted to 45 g, while the daily pollen intake was 20 g.

Some of the workers had been individually marked with differently colored and numbered tags the day prior to the 16 hours of continuous observations (Fig. 2.5). We could thus follow the foraging behavior of individuals. The marked workers spent most of their time in the field

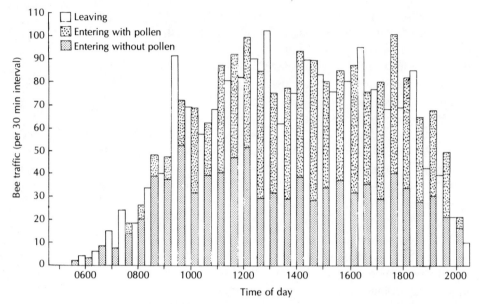

Fig. 2.4 Traffic of bumblebee workers in and out of a large colony (261 workers, 140 new queens, 1,020 eggs, larvae, and pupae) of *Bombus vosnesenskii* near Berkeley, California, on June 6, 1977. Each bar shows traffic over a half-hour period. Many bees stayed out overnight and did not return until early the next morning. (From Allen et al., 1978.)

(Fig. 2.6). Foraging trips were usually one-half to one-and-one-half hours, and bees entering the nesthole in the ground usually reappeared in less than five minutes, even though the tunnel to the nest (which we later dug up) was nearly two meters long. Different bees were bringing in purple, gold, white, lemon-yellow, dark brown, gray, and, sometimes, greenish pollen. The same individuals consistently brought back the same color pollen on consecutive trips. The colony as a whole was tapping the food resources from a wide variety of sources. However, it was not clear if the individuals specialized in different kinds of flowers or if they merely foraged in specific sites having different kinds of flowers (we will examine this question in detail in Chapters 8 through 10).

After determining the flow of traffic and resources in and out of the colony, we dug up the nest and analyzed the contents. The colony contained 261 workers: one old, nearly bald, tattered queen (her baldness was probably due to a virus infection); 136 velvety new queens;

Fig. 2.5 A lightly anesthetized *Bombus vosnesenskii* worker that is being marked with a numbered tag.

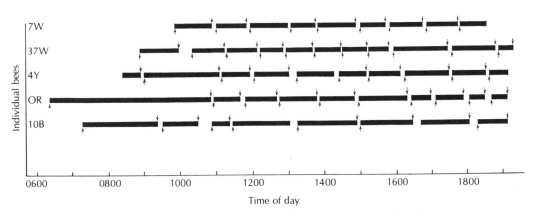

Fig. 2.6 Exits from (↑) and entrances into (↓) the nest by five individually marked *B. vosnesenskii* workers near Berkeley, California, June 6, 1977. (From Allen et al., 1978.)

392 queen pupae; 341 pupae of either drones or workers; 239 larvae; and about 50 eggs. The food stores (at midmorning) consisted of 5.7 g pollen and 260 g sugar, or 195 ml honey (other colonies we dug up contained almost no honey, but numerous full pollen pots). Thus, the honey reserves amounted to six days of net input, and the pollen reserves to 0.3 days. (Incidentally, the honey tasted superb—we all agreed that it tasted superior to any honeybee honey we had ever tasted!) Pollen is important to maintain larval growth, but without sugar to provide energy, adult bees die within hours.

If we assume that the food that the bees had accumulated represented an equilibrium, then it is probable that most of the 50 g sugar and 20 g pollen brought into the colony per day by the worker force of fewer than 260 bees was eaten by the new queens and by queen and drone larvae. Without this drain, the worker force could have conceivably amassed considerable food stores. If honeybees could do as well, then a colony of 40,000 bees should accumulate nearly 15 liters of honey per day. Each bumblebee worker was averaging a net profit to the colony of at least 0.2 g sugar (about 0.3 ml honey) per day. Honeybees have almost no energy drain going to new queens, and only a small drain to drones. Their mode of creating new colonies—swarming—requires that resources be routed into food storage in late summer. The stores are utilized to produce new workers in late winter, so that a single queen can leave the colony with a swarm to aid her in the early spring.

A *B. vosnesenskii* queen weighs an average of 0.43 g, and a drone 0.1 g. If these bees have a pollen-to-bee conversion ratio similar to that observed for some other bees, then one gram of pollen should produce about one gram of adult bee biomass. (However, additional pollen is eaten by bees soon after they emerge from the cocoon.) A forager returning with 0.021 g of pollen per load could supply enough pollen to produce one queen in twenty trips, or one drone in five trips. In the colony we observed, two workers that consistently collected pollen on every trip went on 9 trips per day, like most other workers. This adds up to 0.19 grams per pollen-forager per day under the conditions we observed. Therefore, slightly more than two forager-days would be sufficient to make one queen, and a little less than a day would be sufficient for a drone. The sugar needed for the development of a queen would be about 0.20 g. At 0.025 g sugar per forager-load, this reqirement could be met in eight trips, also a day's effort. But since the bees generally collect both pollen and nectar on the same trip, often full loads of each, this does not constitute an additional foraging day.

We can estimate the foraging effort that must have been mounted by the bumblebees in order to produce the cohort of reproductives found in the nest on June 8. At that time there were 1,160 adult and presumptive reproductives in the nest, including 528 queens (136 adults, plus 392 queen pupae) and 630 drones (341 drone pupae, plus 239 larvae and 50 eggs). This represents a presumptive adult biomass of 290 g ($528 \times 0.43 + 630 \times 0.1$ g). At 0.21 g pollen per forager-day, about 1,380 bee-days of labor would be required to collect the pollen needed to rear the cohort of colony reproductives. These bee-days of labor could be supplied over a long time by a few bees, or over a few days by many. For example, if this labor were to be completed in 30 days (the average egg-to-adult development time), then only forty-six foraging specialists, or about one-sixth of the worker force, would be required. Generally, during the hours when a bumblebee colony is active, about one-third of the colony's population will be away from the nest foraging at any one time. How the bees allocate their labor and decide which tasks to perform is still unknown.

To summarize, the rapid and continuous operation of the bumble-bee factory depends on a dependable supply of energy, and the sole source of energy is the sugar from nectar. As will be shown in Chapters 7 through 11, it is the worker's task to enhance energy flow into the colony by working skillfully and utilizing the best flowers available. Foraging optimization involves achieving the greatest foraging profit for the least cost. The primary energy costs are those of locomotion—shuttling to, from, and among the tiny energy-wells, the flowers. At low temperature, the colony incurs additional operating costs, as the workers must thermoregulate in the nest, as well as outside the nest while foraging. Some bumblebees can forage at temperatures as low as 0°C by stepping up their energy expenditure for heat production in order to elevate body temperature (Chapters 4 through 7). The anatomy and behavior of the bumblebee has evolved to minimize the cost of commuting to and from flowers and working at them. A detailed description of the bees' morphology, physiology, and behavior, in subsequent chapters, will reveal the mechanisms whereby efficiency and energy balance are achieved.

CHAPTER THREE

On glossy wires artistically bent,
He draws himself up to his full extent.
His natty wings with self-assurance perk.
His stinging quarters menacingly work.
Poor egotist, he has no way of knowing
But he's as good as anybody going.

—Robert Frost, *Waspish*

The Flight Machine and Its Temperature

Flight under a wide range of conditions is essential for maintaining the precarious hive economy in temperate regions. However, according to folklore, bumblebees violate aerodynamic theory when they fly. This notion is wrong. It used to be thought that insect flight could be understood on the basis of fixed-wing aerodynamics, when in fact the wings of many insects, including bumblebees, operate more on the principle of helicopter aerodynamics—the action of the wings of bees is essentially like that of reverse-pitch semirotary helicopter blades.

Temperature regulation is crucial to the flight of the bumblebee. Although the idea that bumblebees theoretically can't fly is erroneous, there is some truth in it: bumblebees indeed cannot fly if their muscle temperature drops below 30°C. In order to be active at all and to contribute to the hive economy, a bee must regulate its thoracic temperature, keeping it above 30°C and below 44°C. The flight of the bumblebee is an amazing feat of high-energy work output, but flight and foraging for food, and thus *all* activity in the nest, would be severely limited if the bees lacked the ability to regulate the temperature of their flight motor, the thorax.

Like machines, bumblebees must be supplied with fuel. The only fuel they can use is sugar. As in the internal-combustion engine, the utilization of a given amount of fuel by a bee requires the intake of a given amount of oxygen. Carbon dioxide, water, and heat are produced as by-products. The metabolic rate, or rate of fuel utilization, is a measure of an organism's rate of living. It can theoretically be measured either by what goes into the system or by what comes out. In bees and other insects, it can be conveniently determined by measuring the rate of oxygen consumption; the metabolic rate and the rate of oxygen consumption are therefore generally considered to be synonymous. However, the rates of carbon dioxide and heat emission can also be used to measure metabolic rate. Work output represents only a fraction of the total metabolic rate; metabolism produces considerable heat, as well as useful work.

The metabolic rate of bumblebees in flight is about double, by weight, that of hummingbirds, the most active vertebrate animals. At least 90 percent of the calories expended by a flying bee are released as heat within the thoracic muscles, and the temperature of the thorax of a large bee can rise several degrees Celsius within seconds during vigorous flight. If the bees did not have a cooling system, internal heat production could limit flight at high air temperatures, much as the cruising range and speed of a car can be severely limited if its radiator is defective. In the bees' natural environment, air temperatures fluctuate widely between lows and highs. In temperate bogs, air temperatures often change from near 0°C in the morning to 30°C or more in the afternoon. Since bees require a high rate of work output from their muscles, their activity would be severely limited in time and duration if muscle temperature were not regulated by mechanisms of both active heat retention and active heat loss.

Anatomy is the handmaiden of physiology. The bee's internal anatomy is as beautiful in function as its external form is in appearance. "If you show an engine or a mechanical drawing to a romantic, it is likely he will not see much of interest. The surface reality is dull—lines, numbers. A classical person sees underlying form. Beauty is not what is seen, but what it means" (Robert Pirsig, *Zen and the Art of Motorcycle Maintenance*). Like all insects, bumblebees have three main body parts: head, thorax, and abdomen (Fig. 3.1). (The thorax of the bumblebee contains the first abdominal segment—the propodeum—and the "abdomen" is technically the gaster.) The head has a pair of compound eyes and antennae, both of which are used, among many other functions, in navigation and flight stabilization. These sensors collect information used to correct for yaw, pitch, roll, and variations in speed and power output.

The thorax is the roughly spherical midsection of the body. Attached to it are three pairs of legs and two pairs of wings. The two wings on

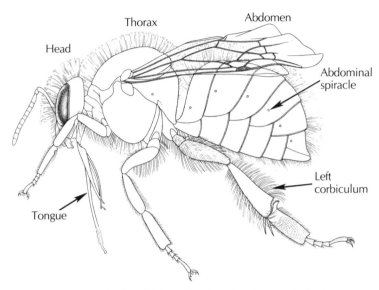

Fig. 3.1 Side view of a bumblebee showing the three main body segments. The full pattern of pile has been shown only on the left hind leg in order to indicate the corbiculum.

each side are joined together by small hooks, so that they act as one during flight.

The wings are small and must beat rapidly—nearly 200 times per second—to keep the bee aloft. The power that is applied to the wings is generated solely by the muscles packing nearly the entire volume of the thorax (Fig. 3.2). There are two main groups of power-producing muscles, each composed of numerous individual muscle fibers (cells) that run the whole length of the muscle (Fig. 3.3). Electron-microscopic photographs clearly reveal the individual muscle fibrils, surrounded by mitochondria and tracheal tubes, within these specialized contractile cells (Fig. 3.4). The mitochondria are the muscles' "power batteries," and the tracheal tubes allow for oxygen and carbon dioxide exchange during the combustion of fuel bathing the fibers.

One set of muscles, the dorsoventral muscles, are attached between the floor and ceiling of the chitinous thoracic box. When these muscles contract, they compress the thorax, and the slight compression is translated by complex wing articulations, acting as levers, into the upstroke of the wings. The downstroke muscles, the dorsolongitudinals, run roughly lengthwise in the center of the thorax and attach to

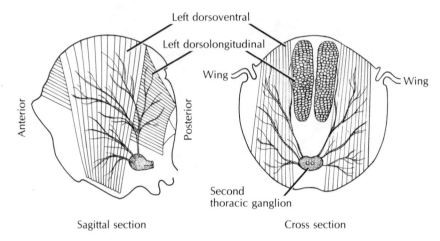

Fig. 3.2 Bumblebee thorax: Sagittal section (taken from slightly left of center) and cross-section, showing individual muscle fibers of dorsolongitudinal and dorsoventral sets of wing muscles, and the nerves that supply them from the thoracic ganglion.

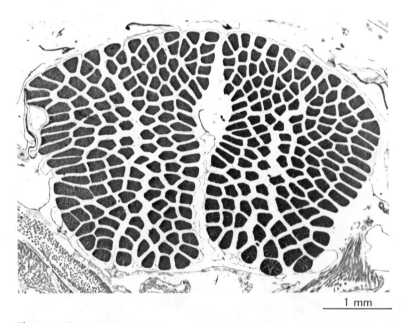

Fig. 3.3 Photomicrograph of cross-section of the right and left dorsolongitudinal (wing depressor) muscles of a bumblebee (*B. vosnesenskii*). A total of 276 muscle fibers are visible. The spaces between the muscle fibers are due to shrinkage, a preparation artifact.

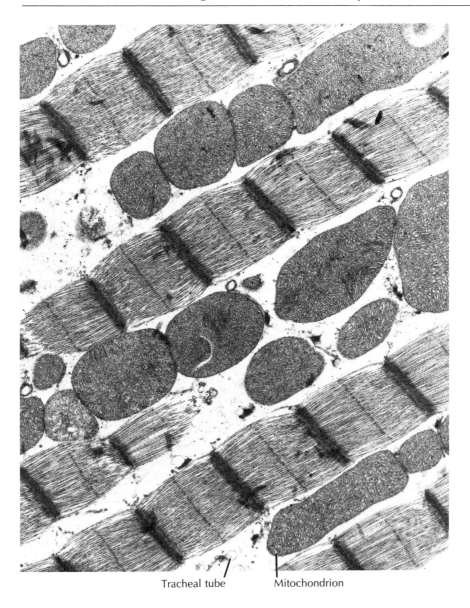

Tracheal tube   Mitochondrion

Fig. 3.4 Electron-micrograph showing bumblebee flight-muscle fibrils surrounded by mitochondria and supplied by tracheal tubes. The mitrochondria provide the muscles with energy and the tracheal tubes allow for air exchange during the combustion of fuel. (Courtesy of M. Ashton.)

projections from the dorsal surface. Their contraction bulges the thorax back to its original position, and creates the downstroke of the wings. Both sets of muscles thus exert power indirectly on the wings, hence they are called indirect muscles. Other muscles—the direct muscles— attach to the base of the wings and regulate their pitch; they affect the bite of the wings into the air and thus, secondarily, the power output of the flight motor.

During flight, the contraction of the dorsolongitudinal muscles causes the dorsoventral set to stretch, and vice versa. In bees, as well as in many other insects with extremely rapid wing-beats, stretching of the indirect muscles is sufficient stimulus in itself to initiate contraction. As a result, very high wing-beat frequencies can be achieved while the muscles are activated by the central nervous system at only 1/10 to 1/20 the actual wing-beat or muscle contraction rate. These indirect muscles are thus described as asynchronous, because contractions are not necessarily synchronous with the receipt of nervous stimuli from the central nervous system, as are all muscle contractions in vertebrate animals. When I flex my biceps muscle, for example, it is being activated by a stream of hundreds or thousands of nervous impulses throughout a single contraction, rather than contracting many times following each nervous stimulus.

The abdomen, or gaster, joins the thorax by a narrow petiole. This slender waist is a characteristic feature of all insects of the order Hymenoptera—bees, wasps, and ants. The abdomen contains a highly distensible honeycrop (Fig. 3.5). When empty, the crop is barely visible. When full, it appears to take up at least 95 percent of the abdominal volume. The honeycrop serves both as a reservoir for food energy for the individual bee and as a bucket for carrying nectar into the nest, where it is emptied by regurgitation.

When not carrying nectar or honey, the bee's abdomen is nearly filled with expanded air sacs. These air sacs are connected to the tracheae, or air tubes, that are part of the ventilatory system found in all insects. In insects, gas exchange is accomplished without the intervention of the circulatory system and without the pair of lungs employed by vertebrates. Oxygen enters via small portholes—spiracles—along the sides of the body, and undissolved gas travels directly in the tracheal tubes to the tissues, where it is used. The tracheae branch into tiny tracheoles that lead directly into the respiring cells. Carbon dioxide moves along the same tubes, but in the opposite direction. The

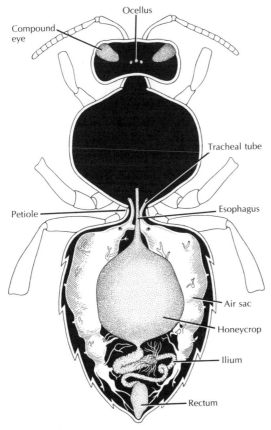

Fig. 3.5 The main body parts of a bumblebee, including the internal anatomy of an abdomen with a moderately full honeycrop.

gases are moved by passive diffusion, by volume changes in the thorax resulting from flight movements, and by active pumping action of the abdomen on the abdominal air sacs, which can act like bellows. Gas exchange is not a limiting factor in flight performance—bees never have a problem getting all the oxygen they need and getting rid of the carbon dioxide. They are never "out of breath," no matter how vigorously they exercise.

The circulatory system is the functional connection between the thorax and abdomen. Among other things, it serves as a fuel line, carrying sugar solution from the honeycrop in the abdomen to the flight motor—the muscles in the thorax—where it is utilized. It does not carry dissolved oxygen and carbon dioxide to and from the respiratory tissues as it does in vertebrate animals. In bumblebees, the blood is pumped from the abdomen into the thorax by a thin tube, the heart, attached under the dorsal surface of the abdomen. The tube traverses the petiole and makes a loop, the aorta, through the thorax. The heart-aortal tube is open at both ends. However, it is not known whether, in bumblebees, blood also enters the heart at holes (ostia) along its length, nor whether blood leaves the aorta before emptying into the head. After blood has left the aorta, it washes freely through the tissues, whereas in vertebrates it is confined in capillaries. The insect circulatory system is therefore called an "open" system. Nevertheless, some of the blood flowing posteriorly in the abdomen is partially confined in a ventral channel, being propelled by undulations of a flap of tissue called the ventral diaphragm.

The nervous system of the bumblebee and other insects consists of a series of ventrally located masses of neurons or ganglia in the abdomen, thorax, and head (Fig. 3.2). The ganglia are connected by paired nerve tracts. The ganglia are the "command" centers that collect information from the sensory organs at the periphery as well as from the interior, integrate this information, "plot" a response, and send commands via nerves to activate muscles or other organs. The largest ganglia are located in the thorax. Thus, decapitation of an insect does not result in immediate cessation of all functions and death, as it does in a vertebrate with most of its integrative centers in the brain.

In the bumblebee, the large muscles that power the wings are innervated by three pairs of nerve fibers from the second thoracic ganglion. One of these nerves supplies neurons to both the dorsolongitudinal and the dorsoventral muscles, and the other two supply only the dorsoventral muscles (see Fig. 3.2). The function of the polyneural innervation of the dorsoventral muscles is unknown.

The insect is enclosed in a chitin shell that hardens within hours after emergence from the cocoon. This external skeleton serves to anchor muscles and limbs, much as the internal skeleton does in the vertebrate body. It is covered with a thin layer of wax that acts as a formidable barrier to water loss. However, once the chitin has hardened, all further growth is precluded; bees with wings (imagos) do not grow after they have emerged from the cocoon. Immature stages are confined in a less restraining armor, which they shed during molting following successive stages of growth.

The performance of both an internal-combustion engine and an insect flight motor is limited by temperature. But the temperature limitation is severe in the insect, while it is of relatively minor consequence in the mechanical counterpart. The insect flight motor must contend with temperature at the beginning of its work, as well as with the heat resulting as a by-product of its operation. Ultimately, all energy utilized is degraded to heat by both the car with the internal-combustion engine (a heat engine) and the insect with its flight motor (a chemical engine). Most of the heat is produced within the motor itself. The upper and lower temperature limits of an internal combustion engine are far apart, the lower limit being set by the ignition system and the upper limit by possible self-combustion of critical components. In the animal, the limits are much closer together and performance is curtailed long before the critical component—muscle—is physically altered, either at the macroscopic or electron-microscopic level. The limits are set by the configurations of complex macromolecules, primarily enzymes, that change shape with changes in temperature. Series of enzymes are required to break down the intermolecular bonds of any fuel, such as sugar, to capture the energy contained in these bonds, and to redirect this energy into useful work, such as muscular contraction.

How does temperature affect an enzyme's action? A fuel molecule passes into a small gap or pocket ("active site") of an enzyme. The enzyme then bends, breaking off a piece of the molecule and releasing the remnants that go to the active site of the next enzyme in the series. Temperature is important in this process for at least two reasons. First, the fuel molecules are in an aqueous solution and the rate at which they randomly move about in this solution and enter the active site of the enzyme is directly related to temperature. Second, and more important, temperature changes cause configurational changes in the complexly folded enzymes. Even slight configurational changes may alter the size and shape of the active site so that the appropriate substrate—the fuel molecule—is not properly bound and released. The analogous situation would be if the carburetor, points, block, and pistons of our car changed shape each time the temperature of the motor changed. The motor could then be tuned to run only at a very narrow range of temperature. This is the case with muscle. The muscles of most animals can produce maximum work output at a sustained rate only in that relatively narrow range of body temperature to which the animal has adapted through evolution, and to which natural selection over millions of years has tuned the physical structure of its biochemical system.

It is not surprising that the muscles of most highly active animals have evolved to operate, and to be regulated, at a high temperature (Heinrich, 1977a). Large, highly active insects, as well as other less active, though larger, animals, heat up from internally produced heat. Their muscles must be able to operate at the temperatures that are internally produced during activity. Those insects that do not heat up and which live at low temperatures have evolved to function at low muscle temperatures. For example, the nonflying grylloblattids (distant relatives of the roaches) that live in the ground at the edge of glaciers are tuned to operate their muscles at about 1°C. They die of overheating in the warmth of a human hand.

It has long been known that bumblebees are capable of seemingly tireless flight, and that they fly at very low as well as at high air temperatures. It has also long been known that flight metabolism must invariably produce heat, which elevates body temperatures. Yet bumblebees were considered to be "cold-blooded" animals, and whether or not they maintained a constant body temperature during constant high levels of activity such as free flight was not known until very recently.

The method used to investigate a problem often determines the specific result, and we tend to accept or reject the result depending on whether or not it fits our hypothesis. That is, we are prejudiced by our methods as well as by our hypotheses. Previous experiments designed to test whether or not large insects maintained a constant thoracic temperature during flight over a range of air temperatures were simple and seemingly elegant; yet they did not answer the question asked. The insect was mounted by the thorax on a light rotating arm of a flight mill. As a result, the animal could fly continuously in small circles, while a tiny temperature sensor measured thoracic temperature. As expected, thoracic temperature immediately began to rise as the insect began to fly. Then, as the rate of heat loss began to equal the rate of heat production, thoracic temperature necessarily stabilized. Because of their thick layer of insulating fur (pile, technically) on the thorax, bumblebees can passively achieve a temperature 65 to 75 percent higher than similarly sized insects having the same metabolic rate. However, high body temperature as such is not proof of temperature regulation. The critical question is this: Is the difference between thoracic and air temperatures during flight greater at low than at high air temperatures? It was not. The unavoidable conclusion from the experiments was that body temperature was not regulated.

I tested the same hypothesis by different methods and got different results. I allowed bees to fly freely in a temperature-controlled room. After they had been in flight for six minutes—sufficient time so that both thoracic and abdominal temperatures had stabilized—they were captured and their body temperatures were measured within a few seconds, before appreciable cooling could occur, with a tiny sharpened thermocouple probed into them. The "grab and stab" technique was simple, and it measured with little error what the animal did normally when it was not encumbered and constrained.

The large queens used for these experiments regulated their thoracic temperatures during continuous flight (Fig. 3.6). As had already been observed in the field, we found that large bees (primarily queens) could remain in free flight even at air temperatures as low as 0°C. In order to do so, they obviously needed to maintain a thoracic temperature at least 30°C higher than the air temperature. The highest air temperature at which they could be induced to remain in continuous free flight in the confines of the room was near 35°C, and at that air temperature they appeared to overheat as thoracic temperature approached the lethal limit of near 45°C. Thus, the minimum difference between thoracic and air temperature during free flight was 10°C, while the largest was about 36°C. The thoracic temperature of the large bees was regulated during free flight at 35°C to 40°C over a range of air temperatures spanning 25°C. Abdominal temperature, on the other hand, was close to air temperature during flight at low air temperatures, and close to thoracic temperature during flight at high temperatures. Small worker bees did not heat up sufficiently to be able to stay in continuous free flight at low air temperatures, but they could fly at higher air temperatures than queens.

Why were the results of free-flight experiments so different from the results of tethered-flight experiments? One simple reason was that heat production was related to flight effort, and unsuspended bees in free flight worked harder than tethered bees. Bees flying while tethered in stationary flight or on the flight mill maintained their "altitude" without any relation to flight effort. A decline in their thoracic temperature had no effect on flight performance, as far as their sensory perceptions were concerned, and hence no compensating reactions were

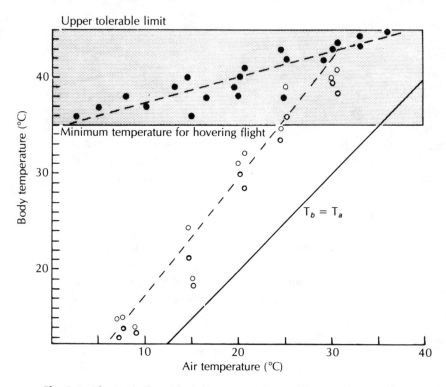

Fig. 3.6 Thoracic (●) and abdominal (○) temperatures of a number of *B. vosnesenskii* queens in continuous flight, each measured at different air temperatures. As air temperature increases, the abdominal temperature also rises, while the thoracic temperature remains between 35°C and 45°C when the air temperature is below 30°C. The solid black line ($T_b = T_a$) indicates what the body temperature would be if it were identical to the air temperature. (From Heinrich, 1975a.)

necessary. They maintained the same flight "altitude" even when their flight effort and their thoracic temperature declined to levels insufficient for normal liftoff and free flight. They were not informed of their "error." In other free-flight experiments, bees were allowed to tank up with measured amounts of syrup so that they had to support greater weights. Loaded-up bees had both higher energy expenditures (work output), as measured by their rates of oxygen consumption, and higher thoracic temperatures than those flying with a nearly empty honey-crop. These increases were directly related to the load held aloft, but unrelated to air temperature, so one could conclude that the rising thoracic temperature with increasing load was not entirely a passive phenomenon. Apparently, the bees regulated a higher thoracic temperature to achieve a higher work output, rather than vice versa. Furthermore, during free flight a bumblebee can, at least in part, regulate its thoracic temperature by monitoring the effects that that temperature has on its flight performance. For example, if a bee is unable to lift a given load with its muscles operating at 35°C, it raises its thoracic temperature by shivering until the muscles are able to generate sufficient work output to keep it aloft. Bumblebees can also monitor their thoracic temperature while they are not in flight, but how they do this and control their body temperature is unknown.

The motivation for flight probably additionally affects thermoregulation. For example, in summer when food is available every day, bumblebee workers usually restrict foraging—and hence flight activity—to air temperatures above 10°C. But they are capable of foraging at air temperatures down to at least 6°C, while queens fly at even lower air temperatures. It is not always clear if the bumblebees' absence from the field at some low temperature is due to inability to thermoregulate, to voluntary relaxation of body temperature control, or to an abundance of flowers so that foraging can be restricted to higher temperatures. Poppy flowers, for example, close up on cool and cloudy days. Drones often stay in torpor on flowers at low temperatures, foraging for their own needs at higher temperatures later in the day, after the workers that forage for the colony economy have been active. The cost-benefit problem of foraging and thermoregulation is discussed in Chapter 7.

Given the opportunity, some bees can regulate their thoracic temperatures behaviorally. In his studies of arctic bumblebees (81° north, at Lake Hazen, Northwest Territories, Canada), Ken W. Richards (1973) found that bumblebees flew at different heights so as to remain in air that was at about 8–10°C. For example, on June 30, 1968, air temperature varied from 15.5°C at ground level to 5°C at 200 cm above ground; the bees flew at an altitude of about 45 cm. But on July 5, air temperatures from ground to 200 cm varied from 24.5°C to 7°C, and the bees, choosing to fly at the same air temperature of 8–10°C, flew about 170 cm above the ground.

The extent to which bumblebees behaviorally thermoregulate in flight in the field has not been investigated in detail. We do know, however, that they can regulate their thoracic temperature to a surpris-

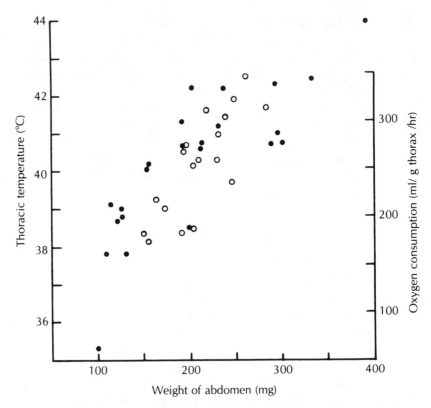

Fig. 3.7 Thoracic temperature (●) and oxygen consumption rate (○) of *B. edwardsii* queens in free flight as a function of the weight of the abdomen, which closely reflects the weight of the sugar syrup carried in the honeycrop. Air temperatures ranged from 10°C to 22°C. (From Heinrich, 1975a.)

ing degree by physiological means. In order to determine whether or not they physiologically regulate, it was necessary to minimize or abolish the possibility for behavioral thermoregulation, by having the bees fly in a small temperature-controlled room of uniform temperature, which allowed them few or no options to regulate their thoracic temperature by other than physiological means.

To understand how bumblebees in the laboratory physiologically· regulated their thoracic temperatures, it was necessary to examine the rates of energy expenditure. The energy expenditure, which is a reflection both of heat production and of work output, was measured by determining how much oxygen the bees removed from a known volume of a respirometer (a 4-liter air-tight cookie jar) in which they had flown uninterruptedly for known durations. Air was withdrawn from the sealed respirometer jar with a syringe before and after flights of known duration, and the oxygen contents of the air samples were measured in an oxygen analyzer. The oxygen consumption rate could then be easily calculated, given the volume of the system, the weight of the bee, and the duration of its flight. The measurements showed that the energy expenditure of free-flying bees, like the changes in thoracic temperature, was unrelated to air temperature, but it did vary as a function of the load the bees held aloft (Fig. 3.7).

It was clear that thoracic temperature during uninterrupted hovering flight was not regulated by increasing the rate of heat production at low air temperatures. Rather, heat production remained the same at 5°C, where the temperature of the thorax was 30°C above ambient, as at 35°C, where thoracic temperature was only 10°C above ambient. If thoracic temperature under the observed conditions were regulated by varying heat production, then the metabolic rate should have been three times higher at the lowest than at the highest air temperatures. It was not significantly different. Therefore, thoracic temperature during free flight must be stabilized by increasing the rate of heat loss at high air temperatures.

In summary, a high temperature of the thoracic flight musculature is both a necessity and a consequence of the bumblebee's flight metabolism. At low air temperatures, thoracic temperature would not be high enough if the bees did not actively retain metabolically produced heat, and at high air temperatures, the bees would overheat during flight if they did not have active heat-dissipating mechanisms. The powerful upstroke and downstroke muscles of the wings each contract about 200 times per second and produce prodigious amounts of heat. Heat production varies as a function of the load carried aloft. But the work effort is generally not varied to affect temperature control. How the bees heat up prior to flight and how they stabilize thoracic temperature once in flight are two additional and entirely different problems.

CHAPTER FOUR

Not all the motion, though, they ever lent,
Not all the miles it may have thought it went,
Have got it one step from the starting place.
—Robert Frost, *The Grindstone*

Warming Up

Social bees could not survive in temperate and arctic regions if they
were not able to generate sufficient heat to elevate both their own
body temperatures and their nest temperatures. Without being able to
elevate their body temperature they would rarely be able to forage,
and they would be unable to heat their nests and speed up the growth
of the young, as is essential for producing and maintaining large hive
populations in the short growing season.

The body temperature of a resting bumblebee, as of other insects, is
generally indistinguishable from the air temperature. To fly, the bee
must raise the temperature of its muscles considerably, so that the
wings can be moved rapidly enough to generate the necessary lift and
thrust for flight. To do this, it shivers.

Many of the larger insects must have a muscle temperature of about
35–40°C—similar to our own body temperature—before they can fly
(Heinrich, 1974c). As in vertebrate animals, there are numerous ways
of warming up and keeping warm. Some butterflies, grasshoppers,
bugs, beetles, flies, and dragonflies warm up by basking in sunshine, as
lizards do. Nocturnal moths, some katydids, and some other beetles,
flies, dragonflies, and bees contract their flight muscles and produce
metabolic heat by shivering. Some insects are capable of both shivering and basking.

The energy of the sun, used for basking, is free. However, it has
drawbacks. First, it is unavailable at night and unreliable in the day-
time. Second, it takes time to bask. Shivering makes the animal inde-

pendent of sunlight and warm temperatures, but the cost is high in energy terms. However, the energy investment for shivering is worthwhile, since without it, the insect would remain limited in its rate of activity and would be grounded at low temperatures.

The metabolic rate at a given muscle temperature, as well as the energy expenditure during a warm-up, can be calculated from records of body-temperature increases in live animals and of the passive rates of cooling in dead animals. Such calculations indicate, for example, that a bee expends approximately 2.9 calories during a warm-up from 24°C to 35°C, 7.5 calories during a warm-up from 13.5°C to 35°C, and 15.7 calories from a warm-up from 6.5°C to 35°C (Heinrich, 1975a). The lower the air temperature, the longer the time of warm-up and the greater the total energy cost of warm-up. The initial rate of thoracic temperature increase, as well as energy expenditure, is low during warm-up at low air temperature, since the cold muscles can work and produce heat only at a low rate. When they are hot they can work at a high rate and produce heat rapidly (Fig. 4.1). Bumblebees generally try to warm up and come out of torpor as rapidly as possible, but the rates of heat production at any one time are a direct function of muscle temperature.

The rapid low-amplitude wing vibrations of a moth or a butterfly getting ready to take off leave no doubt that the animal is contracting its muscles, as vertebrate animals do when they shiver. Bees have similarly been reported to buzz while warming up. Stingless bees, for example, hum during warm-up, and the frequency of the hum increases directly with increasing thoracic temperature (H. Esch, personal communication). Thus, one can determine the body temperature of a stingless bee by listening to the sound it makes. However, bumblebees warm up and keep warm while making no sounds whatsoever and while keeping their wings perfectly motionless. The only externally visible indication of warm-up is abdominal pumping, used to ventilate the rapidly respiring thoracic muscles.

The abdominal pumping movements tell us a lot about what is occurring inside the bumblebee. First, the duration, amplitude, and frequency of the pumping movement indicate the rate of gas exchange for respiration, which is correlated with the rate of energy expenditure. Second, the frequency of the pumping movement itself gives a fairly accurate indication of thoracic temperature (Fig. 4.2). All one needs to

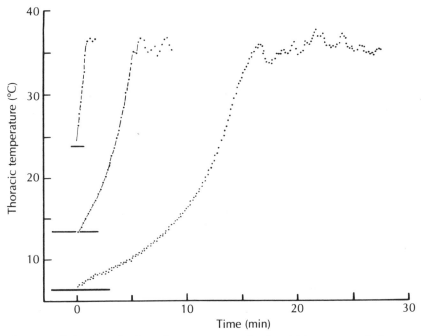

Fig. 4.1 The change in thoracic temperature of a bumblebee queen during three warm-ups from three different ambient temperatures (24°C, 13.5°C, and 6.2°C). Abdominal temperature (not shown) remained low. At low air temperatures, the warm-up takes longer and the rate is lower than at high air temperatures. (From Heinrich, 1975a.)

determine thoracic temperature is a sharp pair of eyes and a stop-watch! At a frequency of 1.1 pumping movements per second, *Bombus edwardsii* queens had a thoracic temperature of 10°C, and at frequencies of 2.2 and 4.4 the temperatures were 20°C and 30°C, respectively. The rate of pumping remains the same for a given temperature regardless of whether the bees are actively warming up, cooling down, or maintaining a stable, elevated thoracic temperature. However, while passively cooling or shivering slightly, bees only pump shallowly, in occasional short bursts, or not at all, and while shivering vigorously and maintaining a thoracic temperature much higher than the air temperature, they pump deeply and continuously. Obviously, high metabolic rates are correlated with abdominal pumping in stationary bees, but this tells us little about the mechanism of heat production.

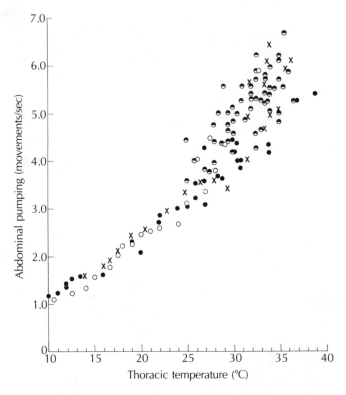

Fig. 4.2 Rate of abdominal pumping of a 0.35 g *B. edwardsii* queen (circles) and a 0.085 g worker (crosses) as a function of thoracic temperature. Measurements were made while the queen's thoracic temperature was increasing (●), decreasing (○), and remaining stable (◐).

The mechanics of shivering in bees could not be examined by direct visual observation of the heat-generating processes that occur inside the thorax. If the muscles in the thorax contract in shivering, they must be activated by nervous stimuli arising from the central nervous system. These signals can be read by electronically eavesdropping either on the nerves directly or on the muscles that they innervate. As in any other kind of "wire tapping," an electrical conductor is used to tap a source, and the signal is relayed to, and analyzed in, a receiver. With the aid of a microscope, we inserted the uninsulated tips of tiny copper wires through pinholes punched into the dorsal surface of the thorax and stuck the wires directly into the flight muscles (Fig. 4.3). The electrical signals were fed into an amplifier. The amplifier, in turn, was connected to an oscilloscope. A movie camera mounted onto the face of the oscilloscope served to give a permanent record of the timing of the electrical events occurring at several muscles from which we could simultaneously record. In addition, electronic counters registered the cumulative number of electrical events (muscle activations). All of the signals were also fed into a tape recorder.

We observed that the bees often activated their flight muscles not only during flight, but also during buzzing and while the wings were held stationary when the bees were "silent" (Fig. 4.4). The flight muscles were always activated during warm-up, sometimes to as high a frequency as during flight itself (Kammer and Heinrich, 1972).

The purpose of the experiment was not only to find out whether flight muscles were activated at times other than in flight. It was also of importance to find out whether, and how, their activity was related to heat production. Concurrent with muscle activation, we also mea-

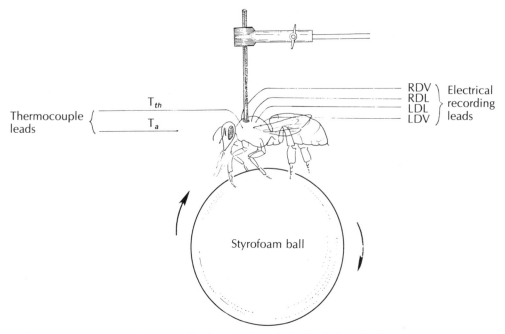

Fig. 4.3 Approximate placement for the insertion of wire leads into the thorax to measure thoracic temperature (T_{th}) and the action potentials from the right dorsoventral (*RDV*), left dorsoventral (*LDV*), right dorsolongitudinal (*RDL*), and left dorsolongitudinal (*LDL*) muscles (see Fig. 3.2). T_a = air temperature.

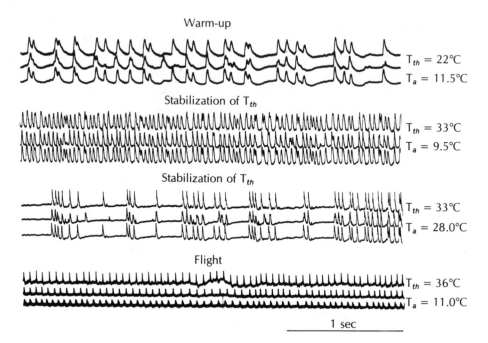

Fig. 4.4 Samples of action potentials recorded simultaneously from three muscle units of the dorsoventral and dorsolongitudinal flight muscles of a bumblebee during warm-up, stabilization of thoracic temperature (while the bee was grounded and stationary) at air temperatures of 9.5°C and at 28.0°C, and during flight. T_{th} = thoracic temperature. T_a = air temperature. (Kammer and Heinrich, unpublished.)

sured the changes in the bee's thoracic temperature and its rate of energy expenditure. Thoracic temperature was measured with a tiny thermistor (electronic thermometer) carefully inserted through a pinhole in the dorsum of the thorax. Another thermistor, placed directly beside the bee, measured air temperature. The bee with its implanted thermistors and electrodes was then sealed into an air-tight respirometer jar, and the air from the jar was circulated by a pump through an oxygen analyzer that measured the oxygen content of the air in the sealed system to an accuracy of .001 percent. The metabolic rate, or heat production, could be computed directly from the amount of oxygen consumed per unit time.

Laboratory experiments have an advantage over field work in that the experimenter can control the variables, so that the effect of one variable at a time can be tested. But there is then no guarantee that the animal will behave in any but the most elementary fashion, if it "behaves" at all. We wanted to give our bee as many behavioral options as possible, so that it could act "normally." At the same time we had to restrict its options in order to control the variables. In order for us to take measurements, the bee had to be fastened and thus could not move from the position where we had placed it. But we did the next best thing by giving it the illusion that it had freedom. First, our bee was fixed in such a way that it could fly in place. (The illusion of flight movement at varying speeds can be given by presenting the bee with a moving visual field, such as a rotating drum of moving stripes. Rather than letting the bee move over the stationary visual field, one can let the visual field move under the bee.) Also, the bee could walk as "far" or as fast as it wanted. This illusion was created by letting it grasp a light styrofoam ball. As it walked on the ball, the ball rotated, and the bee, being dorsally attached, remained at the same spot—"on the ball," that is. We controlled air temperature about the bee by dipping the respirometer containing the bee into a temperature-controlled water bath.

In summary, we could not trail a bee with a laboratory attached to it while it was foraging on hot or cool days in a field of goldenrod, but we did the next best thing and looked at a few components of the foraging behavior, such as flight and heat production, which we ultimately hoped to relate to the bee's untrammeled behavior in the field. Bees in the respirometer told us that increases in thoracic temperature were always correlated with activation of the thoracic muscles, as in flight (Fig. 4.5). Metabolic rate was directly related to the number of depolarizations (activations) of the flight muscles. Furthermore, the energy expenditure associated with a given rate of muscle activation was similar whether the animals were in flight or whether they were stationary and had their wings folded over the abdomen (Heinrich and Kammer, 1973). Walking, with no or very little activation of flight muscles, required negligible energy expenditure in comparison to vigorous shivering or flight. Walking bees did not heat up from walking per se. The results were clear: the bees produced heat in the thorax by shivering, utilizing the flight muscles, which were "uncoupled" from the wings. The faster the rate of warm-up, and the greater the difference in temperature the bees maintained between the thorax and the air, the greater the action potential frequency and the metabolic rate.

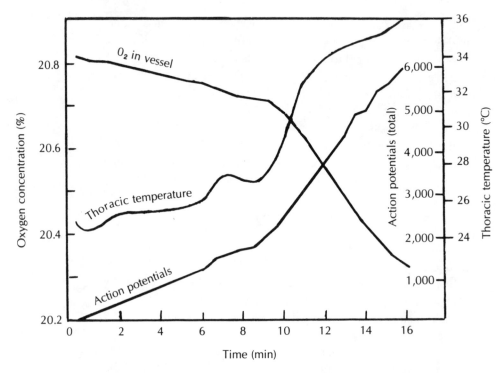

Fig. 4.5 The thoracic temperature increases concurrently with decreases in the oxygen concentration in an experimental vessel (volume = 250 ml), and with increasing numbers of action potentials recorded from the thoracic fibrillar muscles in a *B. vosnesenskii* worker. (From Kammer and Heinrich, 1974.)

We do not know precisely how the various muscles and their sub-units behave relative to one another while a bee shivers. We do know that shivering involves contractions of the main power-producing flight muscles, following their activation by the central nervous system. In bumblebees, these muscles are extensively innervated—each muscle cell has neuronal connections. Possibly, groups of muscle cells, which are arranged into numerous packets (see Figs. 3.2, 3.3), are activated in relays rather than all at once, thus creating a steady tension of the whole muscle mass with little or no overall movement of the whole muscle or the wings connected to them. Isolated bumblebee asynchronous flight muscles are known to contract directly in a one-to-one manner in response to nervous stimulation, like the synchronous muscles of vertebrate animals (Ikeda and Boettiger, 1965). However, it is most likely that the muscles, while being mechanically quite active during warm-up, cause no wing movements for quite another reason. A small muscle, the third axillary muscle, when relaxed unhinges the wings from the main power-producing muscles, whose contractions produce most of the heat. Thus, when the pleurosternal muscle, acting like a clutch, is relaxed, the main power-producing muscles of the wings can be working without moving the wings. This warm-up mechanism is used not only in flight preparation, but also in nest temperature regulation.

CHAPTER FIVE

Burly, dozing humble-bee
Where thou art is clime for me
Let them sail for Porto Rique,
Far-off heats through seas to seek,
I will follow thee alone,
Thou animated torrid-zone.

—Emerson, *The Humble-bee*

Heating the Nest

The nests of social insects are factories that produce babies. The larvae, being very sensitive to temperature extremes, must be kept warm. Drops in temperature below 30°C stunt their growth or may result in developmental defects such as wrinkled wings in some bee species. In order to build a defect-free product in large numbers and in a short time, bumblebees must invest significant portions of their energy supplies in maintaining a constant temperature in their nests, particularly when they are living in cool temperate and Arctic regions. In any area, however, considerable energy savings are possible through appropriate nest placement.

Many species of bumblebees place their nests inside tree cavities or rodent burrows, where temperature fluctuations are buffered. Arctic bumblebees nest on top of the ground in lemming nests, thus avoiding the permafrost and making use of solar heating at the ground surface. In addition, by nesting inside bird or rodent nest cavities, bumblebees are taking advantage of existing insulation to retain metabolically produced heat.

Ground-nesting species supplement the existing nest insulation by incorporating bits of grass and plant fibers into the nest. They will use almost any type of insulation available, as illustrated in a colony of *B. occidentalis* we maintained in a bare box in the laboratory. We wadded a roll of wet, thick, blue filter paper into a plastic tube to maintain nest humidity one day. To our surprise, by the next day the entire nest was covered with a blue canopy—the bees had pulled the filter paper from the tube, shredded it, dried it, and dragged it to the nest, where they had anchored it with wax into a solid canopy.

Bees keep the materials covering the nest surface dry and fluffy, thus enhancing their insulative properties. In addition, bumblebees usually build a waxen canopy, as just described, over the nest. This cover traps the rising warm air resulting from the bees' activity in the nest. Holes in the canopy are repaired in minutes. In populous bumblebee colonies, the metabolic heat produced by the many bees is often sufficient to warm all of the air in the nest, as well as the brood, without much incubation. This is advantageous, since incubating bees are immobile and unavailable for other hive duties.

Nest insulation is varied to suit conditions. When the nest is in danger of overheating, the bees partially remove the wax canopy and circulate air through the nest by fanning. Nest insulation should be of particular importance in the early stages of colony founding, when the queen must be absent from the nest to go foraging. But so far, no systematic studies have been made of the insulation in bumblebee nests.

Honeybees, like bumblebees, also regulate their nest temperature. They employ fanning and evaporative cooling at high temperatures and energy-intensive metabolic heat production at low temperatures. A major requirement for successful overwintering is a large energy supply, in the form of honey, for heating. But the bees minimize the cost of this energy by only heating selected portions of the nest—they concentrate on the area with brood. When there is no brood, as in midwinter, they crowd into a tight cluster close to the honey stores. Small or loose clusters of bees are unable to withstand the low temperatures of deep winters. In contrast to European honeybees, the African honeybee has limited ability to form and maintain such a tight cluster, and hence it is thought that this bee (the so-called "killer" bee), introduced into South America, will not invade the northern regions of North America. In winter, bees leaving the safety of the cluster usually die within a few minutes. Unlike bumblebees and many other bees and other insects that hibernate in winter, the tissues of honeybees have not evolved the ability to withstand freezing temperatures.

Bumblebee queens emerging from hibernation in the spring face the task of regulating not only their own body temperature, but also the temperature of the nest they initiate. It has long been known that established bumblebee colonies with many workers maintain a relatively constant nest temperature near 30°C, despite major fluctuations in air

Fig. 5.1 A captive *B. vosnesenskii* queen incubating her brood clump. The brood consists of pupae from her first batch of eggs, plus a second batch of eggs that is encased in the waxen envelope underneath her right leg. The bee is facing her honeypot. At air temperatures ranging from 3 °C to 33 °C, the thorax was 35–38 °C, the abdomen was 31–36 °C, the brood was 24–34 °C, while the honeypot remained at about the air temperature. (From Heinrich, 1974c.)

temperature, by producing heat at low temperatures and by fanning at high (Hasselrot, 1960). Until recently, however, little was known about the role of individual bees in temperature regulation.

A queen founding her colony perches upon her brood clump both day and night when not foraging, and she assumes a characteristic posture. She extends her legs and elongates her abdomen, wrapping herself about the brood clump while facing the honeypot, usually sipping its entire contents in the course of a single night (Fig. 5.1). Like the naked brood patch in birds, the major area of contact between the bee and the brood clump is the ventral surface of the abdomen, which is the only relatively large body area nearly devoid of hair. The thorax and the dorsal surface of the abdomen are covered with a dense layer of pile, which nearly halves the rate of heat loss from the covered body surface.

Under ideal conditions the eggs in the incubated pollen clump hatch in 3 to 4 days. The larvae feed on the pollen, grow rapidly, and pupate in as little as 10 days. In another 11 days, the adults may emerge. However, egg, larval, and pupal development times vary greatly between colonies. Development stops or greatly slows when the queen runs low on food and stops incubating.

Several biologists in the nineteenth century presumed that the bumblebee queen incubates the pollen clump containing eggs, larvae, and pupae. In 1837 the British surgeon George Newport found from his pioneering measurements of the body temperature of beetles, moths, and bees—using specially designed mercury thermometers "scarcely larger than crow quills"—that some bumblebees, the "Nurse Bees" (both workers and drones), "seem to be occupied almost solely in increasing the heat of the nest and communicating warmth to the nymphs in the cells by crowding upon them and clinging to them closely." These observations confirmed what had already been supposed 35 years earlier in France by Pierre Huber, one of the first students of bumblebees.

Newport's account is interesting for other reasons—for example, his comparison, in the same publication, of the blood circulation in the dura mater of the brain of a patient he had trepanned with the blood circulation in moths and caterpillars. However, despite his imperious breadth of knowledge and experience, his observations on bumblebees lose some of their credence when we read that he observed "profuse perspiration" in insects and that "incubating bumblebees gradually become bathed in perspiration because of their great work." (Insects can become hot, but they have no sweat glands.) Newport reported a body temperature in a "very excited" *B. terrestris* worker of 34.5°C, at an air temperature of 23°C, as measured by an external application of his thermometer.

The issue of incubating bees was raised again nearly half a century later, in 1882, by the Austrian entomologist Eduard Hoffer. Hoffer stated: "While incubating, the queen frequently lies on the egg-cell in such a way that it warms it with its abdomen, as a hen does her eggs, the abdomen being closely pressed against the cell. Moreover, she also resorts to brooding in the case of older clumps of eggs and larvae, and the cocoons." These interpretations of the queen's behavior were later rejected by the German entomologist H. von Buttel-Reepen, who proposed a more "reasonable" hypothesis: that, rather than incubating, the bumblebees were warming themselves with the heat given off by the larvae.

I measured brood-clump temperatures with and without an attending queen and found that the queen was indeed, like a hen, incubating her brood (Heinrich, 1974a). Thermocouples imbedded in the brood and connected to a recording device showed that the temperature of a brood clump containing large larvae was close to air temperature in the absence of the attending queen (Fig. 5.2). But when the queen perched on the brood clump, the temperature of the clump immediately increased. When the queen left the brood clump, its temperature immediately declined until it became nearly identical to air temperature. Clearly, the bee was heating the brood, and not vice versa. As long as a queen had ample sugar syrup available, she spent most of her time, day and night, incubating. Heat transfer from bee to brood was direct and efficient. Bees could maintain a brood clump as much as 25°C above air temperature even in the absence of nest insulation.

Bumblebee queens, at or near the time that they initiated nest construction, maintained a high thoracic temperature almost continuously (Heinrich, 1972c), even when apparently not intending to fly (Fig. 5.3). But their patterns of body temperature while incubating were different from those of bees in flight. During incubation, both thoracic and abdominal temperatures were regulated and maintained relatively independent from air temperature (Fig. 5.4), while during flight, abdominal temperature was not regulated and remained low except at high air temperatures. During flight, the difference between abdominal and air temperature increased with increasing air temperature, whereas the opposite relationship held during incubation. Thoracic temperatures of incubating bees ranged from 34.5°C to 37.5°C at air temperatures from 3°C to 33°C, while the temperature of the abdomen was regulated at only a slightly lower temperature than that of the thorax. As discussed in detail in the next chapter, the abdomen is heated by variable heat flow from the "heat engine," the thorax. During foraging at low temperatures, this heat flow was much reduced (Heinrich, 1972a), minimizing the loss of calories to the abdomen, and during incubation it

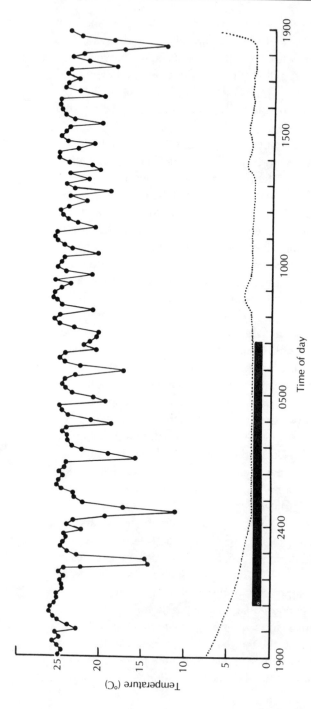

Fig. 5.2 Temperatures measured near the center of a brood clump for one day with a *B. vosnesenskii* queen in attendance, but with nest materials lacking. Temperatures were recorded at 11-second intervals, and the averages for 10-minute intervals are shown. The dotted line represents air temperature. The brood temperature dropped conspicuously when the queen left the brood temporarily in order to feed. Unlimited quantities of honey were available nearby. (From Heinrich, 1974a.)

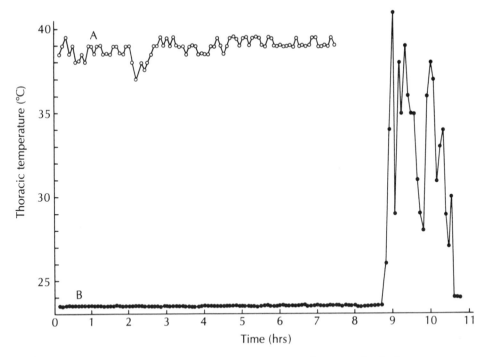

Fig. 5.3 Thoracic temperatures of (A) a captive *B. vosnesenskii* queen that had initiated nest building and maintained a high thoracic temperature continuously at night (7 P.M. to 3 A.M.) while remaining relatively stationary and (B) an overwintering *B. vosnesenskii* (from 1 A.M. to 12 P.M.) that had not yet initiated nest building. Temperatures were recorded at least once per minute and are here plotted at 5-minute intervals. Ambient temperature was 23–25°C. (Adapted from Heinrich, 1974a.)

was greatly accelerated. Since the temperature of the brood was maintained relatively independent of the air temperature, it is apparent that the bees were imparting more heat to the brood at low than at high air temperatures.

The rate of temperature increase of the brood at the start of incubation, as well as the final brood temperature, was directly related to the difference in temperature between the abdomen and the brood. Since the bees regulate the temperature of the abdomen, we can assume that they replace heat into it as soon as heat is lost to the brood or elsewhere. The heating rates of the brood, as well as the final temperature

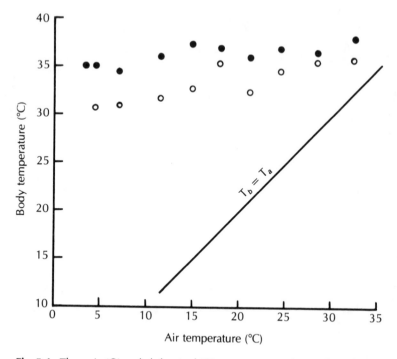

Fig. 5.4 Thoracic (●) and abdominal (○) temperatures of a number of incubating *B. vosnesenskii* queens at different air temperatures. Both abdomial and thoracic temperatures were regulated by the bees during incubation, while during flight abdominal temperature was not regulated. The line $T_b = T_a$ indicates what body temperature would be if it were a passive function of air temperature. (From Heinrich, 1974a.)

of the brood, are explicable in terms of the regulation of abdominal temperature alone.

An experiment showed that heat flowed into the brood primarily by way of the abdomen rather than directly from the thorax. A dead bee with an electrical resistor implanted either into the thorax or into the abdomen was fixed onto a brood clump in the incubation posture (Heinrich, 1972d). Temperatures were simultaneously recorded from thermocouples implanted in the thorax, the abdomen, and the brood. The amount of heat input to the resistor could be precisely controlled by varying the current input with a transformer. Heating the thorax of the dead bee had little effect on either abdominal or brood temperature. But heating the abdomen with the resistor caused a parallel, though slightly lower, increase in brood temperature. The conclusion was that the brood was heated by way of the abdomen.

Honeybees also regulate the temperature of their brood to about 30–33°C, but they do not incubate. At low temperatures, the bees crowd onto the brood comb, nearly covering it, like a blanket, and producing heat by shivering, yet these bees are free to move about and perform other hive duties.

The bumblebee queen deposits a scent (pheromone) where she has oviposited (Heinrich, 1974b). This reminds her where her eggs are located, and so she heats only the small localized area of the nest that contains the young, rather than heating the whole nest chamber. By means of this scent, she can locate her young, which are often completely covered with wax, even in the darkness of the hive.

The European hornet, *Vespa crabro*, and related paper wasps also incubate their brood. Interestingly, in wasps only the pupae, which do not feed, are incubated. The larvae, which are fed directly in the open combs, use the food they receive from their colony-mates as fuel for heat production (Ishay and Ruttner, 1971).

What is the energy cost of brood incubation? For an incubating queen in the laboratory, the metabolic rate (the energy expense of heating and keeping warm) was a function of air temperature, much as it is in any warm-blooded animal. However, the metabolic rate of an incubating queen was much higher than that of the proverbial shrew that must eat several times its own weight per day to stay alive. At low air temperature, the maximum metabolic rate was almost as high as during free flight (Fig. 5.5). But the bees in the laboratory had an unlimited food supply and they often incubated day and night, pausing only long enough to refill at the feeder provided for them. In the field, bees would presumably have to spend much more time foraging and have less time available for incubation.

How much does a bee have to forage to secure the food needed to provide energy for incubation? The answer to that question depends a great deal on the kinds of flowers available and their nectar contents. In the early spring when colonies were being initiated, queens were foraging from blueberry flowers, which contained on the average 0.04 mg sugar, or the equivalent of about 0.15 calories. While incubating uninsulated nests in the laboratory at 5°C, each bee consumed approximately 80 ml oxygen per gram of body weight per hour. For a bee weighing one-half gram, this is equivalent to an energy expenditure of

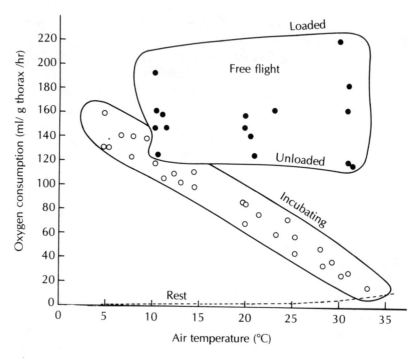

Fig. 5.5 The rates of oxygen consumption of *B. vosnesenskii* queens as a function of air temperature during continuous free flight (○), while incubating the brood (●), and while at rest (---) (no flight muscle activity). (From Heinrich, 1975a, and Kammer and Heinrich, 1974.)

75 calories on a carbohydrate diet, which is equivalent to the sugar content of 500 blueberry flowers. At one blueberry bush, each queen visited approximately 15 flowers per minute. These figures suggest a minimum foraging time of 33 minutes at a rich food source to supply the fuel for an hour's brood incubation at 5°C. Obviously, unless their nests were highly insulated, the bees in the field could not incubate for the long durations that were observed in the laboratory, where the food supply was immediately available to them in unlimited amounts.

The maintenance of a high body temperature is considerably more costly on a weight-relative basis in a small than in a large animal. It has been found empirically that for animals the equation $0.031(\text{kg body mass})^{-0.51}$ describes the metabolic cost of keeping warm in terms of milliliters of oxygen per gram per hour per °C elevation of body temperature. Although this is an approximation, and can be used only for rough estimates, it does indicate that a large queen bee weighing 0.0005 kg would expend, per unit body mass, about five hundred times more energy than a human weighing 73 kg to maintain the same elevation in body temperature. The bees' maintenance of a high body temperature is thus impressive, particularly when we consider that bumblebees can exist in the Arctic where actual temperatures are often low to begin with and where they regulate not only their own body temperatures, but also the temperature of nests.

Kenneth Richards (1973) has measured the nest temperatures and activities of the Arctic bumblebee *B. polaris* at Lake Hazen (81° north) on Ellesmere Island, Northwest Territories, Canada. At the initiation of the colony cycle, the lone queen maintained a nest temperature of about 25–30°C when she was in the nest. However, she made foraging trips at frequent intervals, and when she left the nest its temperature declined. But, at an air temperature of 10°C, the nest temperature generally did not decline more than 7°C in the half hour or so while the queen was foraging. Thus, unless the larvae of these Arctic bees are endothermic, which is unlikely, it is probable that the nests are highly insulated, which must be an important factor in the bees' energy economy in the harsh polar environment. After all of the sixteen to seventeen larvae of the first brood have developed into workers, nest temperature is maintained steadily near 35°C, and the comings and goings of the queen, or of a few foragers, then have no noticeable effect on nest temperature. (Fig. 5.6). It is of interest that Arctic bumblebees appear to have higher nest temperatures than honeybees and bumblebees from temperate regions. This needs to be studied in more detail, but it is tempting to speculate that the higher nest temperatures of the Arctic bumblebees, as well as the larger initial brood, are adaptations to speed up the colony cycle in the very short growing season.

Given very adverse conditions, how long can a bee incubate if she has maximum fuel reserves? At 2° to 3°C, a captive queen with an uninsulated brood clump fed every 1 to 2 hours. The moderately filled honeycrop of a queen contains approximately 0.2 ml of nectar. If the nectar has a sugar concentration of 30 percent, the bee can carry 240 food calories. Utilizing 75 calories per hour (at 5°C), the bee should metabolize the food in her honeycrop in 3.2 hours. Since the honeypot contains an additional 0.5 ml of nectar, the bee can incubate uninterruptedly for about 11 hours at 5°C if she utilizes all of the food reserves stored in her honeycrop and honeypot. In the field, the bee's metabolic

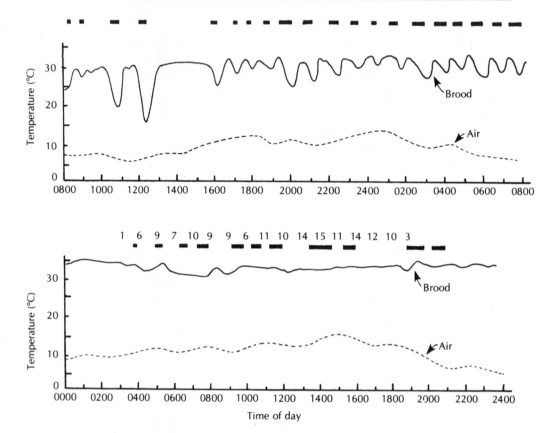

Fig. 5.6 Top graph shows relationship between brood temperature and foraging activity (black horizontal bars) of a *Bombus polaris* queen in the field over a
24-hour period, June 22–23, 1968, at Lake Hazen, Northwest Territories, 88
km south of the North Pole. At that time the colony consisted only of the queen
and seventeen larvae. On July 6–7 (bottom graph), the colony contained sixteen workers and a second brood of ten eggs, in addition to the queen. Although both the queen and workers left the nest to go foraging, brood temperature was more stable. Numerals indicate the number of hourly foraging trips by
workers. (From Richards, 1973.)

expenditure would seldom have to be this high, because nests are always insulated. The food reserves gathered in the day are generally
sufficient for only one night. When the bees exhaust their food, they
enter torpor and cease to incubate. F. W. L. Sladen, who published a
monograph on British bumblebees in 1912, observed: "Occasional

periods of semi-starvation, lasting for a day or two, do not harm a colony of bumblebees: the bees simply become drowsy, remaining in a state of suspended animation."

Temperature regulation appears to require an exorbitant number of food calories. As a result, huge numbers of food calories are used to produce vanishingly few calories of living matter. In most mechanical systems, we tend to think of heat as waste—a sometimes unavoidable by-product that we go to great lengths to minimize, for the sake of energy economy. Many cold-blooded animals are notoriously good at converting a large percentage of their food calories into babies. Bumblebees convert only a tiny fraction of their ingested calories into babies. We may ask: Are the bees "inefficient"? Why are they not outstripped by those that have better food-to-babies conversion ratios?

Evolution acts to increase output of a product by increasing the efficiency of morphology, physiology, and behavior to minimize the input needed for that product. However, the name of the game is not efficiency, per se, but output of successfully reproducing offspring, period. And sometimes production must take place at extreme environmental conditions or in the face of competition, so that high operating costs are unavoidable.

There are times when resources are scarce, but otherwise dependable, and efficiency in their use determines life or death. There are other times when resources are amply, but only temporarily, available, and competition and reproductive advantage will be decided not on the basis of efficiency alone, but on the speed of resource acquisition, even if it involves waste. Nectar and pollen are very concentrated food resources that are available only for a limited amount of time, both seasonally and diurnally. And bumblebees have only a short time in which to build up the colony. Elevation of body temperature in adults, in and out of the nest, accelerates larval growth rates and the speed of resource harvesting. We are thus analyzing not only efficiencies of food conversion, but also energy investment and procurement strategies for living and reproducing within a limited time span at given resource distributions. Bumblebees get all of their energy from the nectar of flowers. This energy can be converted into time, in the sense that it is used to resist torpor, to promote activity, and to increase the number of meaningful events that constitute the biological (rather than the absolute) scale of time.

Nevertheless, efficient use of calories acquired is a prime requisite for reproductive success. The rate of fuel utilization is directly translatable into the rate of heat production, which in turn can be translated into the growth rate of immature bees. The less fuel available, the less the young can be incubated, and more time will be required to bring them to adulthood. In view of the high energy costs, it is not surprising that evolution would perfect the efficiency of transfer of heat from the muscles in the thorax, where it is produced, to the brood, where it is needed. This heat transfer involves two steps: postural adjustments of the body to the brood clump, and physiological mechanisms for heat transfer from the thorax to abdomen.

CHAPTER SIX

And all our blandishments would seem defied,
We have ideas yet that we haven't tried.

—Robert Frost, *Riders*

The Heat Transfer System

The greater the size of the engine, the less rapid the loss of the internally generated heat during the high energy-expenditure of locomotion. That is why a Lincoln Continental must have a cooling system of circulating fluid to dissipate heat from the motor via the radiator, while the much smaller Volkswagen Beetle can be air-cooled without any special heat-transfer mechanism. The same principles apply to insects. Some insects that are large and vigorous flyers—sphinx moths, certain dragonflies (Heinrich and Casey, 1978), and bumblebees—could overheat in flight if they did not use their abdomens as heat radiators. Small insects, and those that are less powerful flyers, are cooled passively by convection as they move through the air, and they do not need to assist this process by transporting the internally generated heat to the exterior.

Some experiments described previously showed indirectly that heat transfer was under physiological control. During free flight at different air temperatures, the rate of heat production in bumblebees was constant. By inference, active heat loss was necessarily directly proportional to air temperature, in order for the observed thoracic temperature stabilization to occur. But during incubation, the rate of heat transfer from thorax to abdomen had to be nearly maximal at all times, in order for the abdominal temperature to remain nearly as high as the thoracic temperature.

Incubating bees were found to regulate abdominal temperature near 35°C, about 2°C lower than thoracic temperature. Furthermore, no in-

crease in abdominal temperature was ever observed without the thorax having first been heated, and increases of abdominal temperature
were often correlated with decreases of thoracic temperature. Abdominal temperature was never higher than thoracic temperature. Since the
heat-producing machinery is the musculature in the thorax, it was apparent that the bees possess an efficient mechanism for transferring
heat between thorax and abdomen, one that functions not only during
thoracic temperature stabilization in flight, but also during brood incubation. What kind of biological engineering made the heat transfer
possible?

A mechanic can often deduce how a machine runs from the way it is
constructed. Similarly, anatomy reveals much about function. A close
examination of the bee's anatomy gives clues to the heat-transfer
mechanisms that can be tested by physiological experiments. Since
previous work with sphinx moths had shown that these nearly continuous fliers have a fluid-cooled "flight engine," with the circulating
blood used for cooling (Heinrich, 1971; Heinrich and Bartholomew,
1972), it seemed that the circulatory system might be a good place to
start looking for the heat-transfer mechanism of the bumblebee.

The bumblebee's circulatory system turns out to have some peculiar
differences from that of the sphinx moth. In both insects, the heart is a
tube lying close under the dorsal surface of the abdominal wall. In the
bumblebee, however, the heart makes a large ventral loop beneath an
air sac in the anterior portion of the abdomen (Fig. 6.1). The air sac
should help to insulate the thorax from the abdomen. But what is the
function of the loop of the heart? The loop brings the cool blood entering the thorax via the heart into close physical proximity with the
warm blood leaving the thorax (Figs. 6.2, 6.3). Blood flows posteriorly
underneath the ventral diaphragm that lies in contact with the heart-
loop, as well as in spaces directly surrounding the loop. Blood flows
anteriorly through the loop. In such an arrangement, heat from the
warm blood leaving the thorax would be partly absorbed by the cool
blood entering via the loop. Thus, the anatomical arrangement appears
to be designed to retain heat in the thorax or retard its flow from the
thorax to the abdomen (Heinrich, 1976e).

Heat transfer between oppositely flowing blood is possible because
fluid flow is confined by channels while heat flow is not so limited.
Heat will always flow from a higher to a lower temperature. Hot blood
entering the abdomen around and close to the loop comes within a

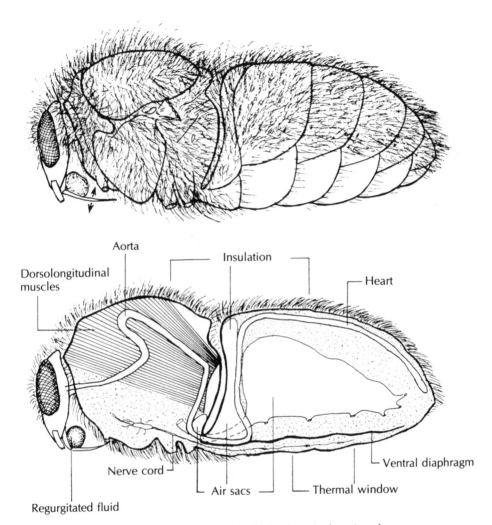

Fig. 6.1 Diagrammatic representation of a bumblebee's sagittal section shows features that act to retain heat in the thorax as well as to transfer heat from it. The thorax and dorsum of the abdomen are heavily insulated with pile, but the venter of the abdomen is relatively free of pile and acts as a thermal window. The narrow petiole between thorax and abdomen and the air sacs at the anterior portion of the abdomen act to retard heat flow to the abdomen. Cool blood is pumped anteriorly in the heart where it is warmed in the aorta that passes between the right and left dorsolongitudinal muscles. Undulations of the ventral diaphragm propel warm blood posteriorly. (From Heinrich, 1976e.)

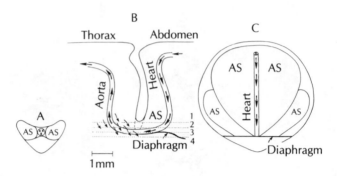

Fig. 6.2 Diagrammatic representations of cross-section of petiole area looking anteriorly (A), of sagittal section of petiole area (B), and of cross-section anterior of abdomen looking anteriorly (C). Arrows show direction of blood flow. AS indicates air sacs or tracheal tubes (in the petiole). Horizontal lines (1, 2, 3, 4) indicate area of cut shown on photomicrographs in Figure 6.3. (From Heinrich, 1976.)

few (probably two) cell layers of the cool blood entering the thorax. As long as blood is simultaneously flowing in both directions through the petiole and anterior abdominal area, some of the heat in the blood entering the abdomen should flow into the blood returning to the thorax. This flow of heat, through blood vessel walls from one blood stream into another blood stream flowing in the opposite direction, is called countercurrent exchange.

Countercurrent systems are well known in vertebrate animals. For example, the flippers of seals, the tails of beavers and of muskrats, and the legs of many shore birds have countercurrent heat exchangers that allow these animals to be active in ice water while minimizing the heat flow to their limbs, and thus retarding energy leakage to the environment. In these animals, the cooled blood returning to the body from the terminal portions of the limbs can be shunted into many small veins—the countercurrent exchanger—that form a sheath around the arteries. This venous blood picks up heat from the arterial blood before it travels very far down the limb. Heat loss, on the other hand, is accomplished by bypassing the small interior veins; the blood is shunted instead into the surface vessels of the extremity. A beaver that is prevented from using the countercurrent exchanger in its tail must shiver almost continuously to keep from freezing to death while it is in icy water. That is a high price to pay for keeping its tail warm.

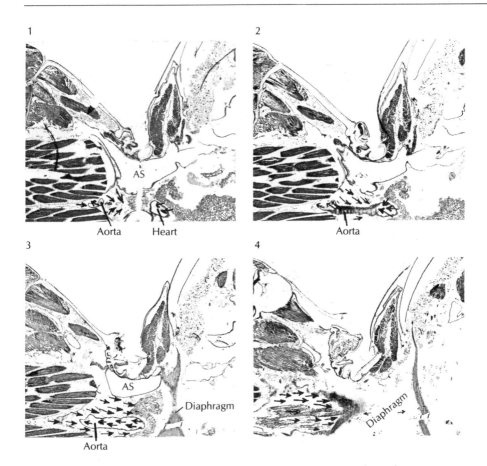

Fig. 6.3 Photomicrographs of petiole area taken from locations indicated in Figure 6.2, showing sections from the right side. Arrows have been superimposed to indicate directions of warm (to right) and cool (to left) blood flow. Note (in 2 and 3) that the system acts as a countercurrent heat exchanger: when blood is flowing simultaneously in both directions in adjacent channels, some of the heat being carried posteriorly into the abdomen is recovered by the cool blood returning to the thorax.

In bumblebees, heat loss from the thorax to the abdomen by blood returning to the abdomen would also be great without countercurrent heat exchange, since continual blood circulation is necessary to carry nutrients from the abdomen to the working thoracic muscles. Without heat-retention mechanisms, the circulating blood in the thorax could cool and cause cooling of the muscles, so that flight and the high rate of heat production could no longer continue. During flight at 5°C with a thoracic temperature of 35°C, for example, the circulation of only 0.5 ml of fluid per minute could potentially remove 0.8 of the 1.2 calories produced per minute, and cause the thoracic temperature to plummet. However, the anatomical arrangement of the circulatory system in the petiole and abdomen allows heat leaking into the abdomen by way of the blood to be at least partially recovered by the blood entering the thorax. Two other anatomical features also help conserve heat in the thorax: (1) the thick layer of pile on the thorax, and (2) the insulating air sac on the anterior inside surface of the abdomen.

In contrast, other anatomical features of the bumblebee appear to be designed to increase heat loss. First, as in sphinx moths and some other large bees (*Xylocopa, Euglossa, Xenoglossa*), but unlike most other insects, including honeybees, the aorta (the continuation of the heart into the thorax) makes a large loop through the flight musculature. Blood flowing through the coil-like bend of the aorta must necessarily pick up heat from the working muscles. If this heat were not recovered by the countercurrent heat exchanger, it would be transferred with the blood into the venter, the ventral portion of the abdomen. The venter is lightly insulated, so that heat traveling into the abdomen should be lost by convection to the air, or by conduction to the brood.

It seems improbable that an insect would have two apparently antagonistic anatomical designs, unless it also has mechanisms to use them alternatively, or selectively. How are they used? Countercurrent heat exchange in the bumblebee would be reduced or eliminated, and heat flow into the abdomen greatly accelerated, if the blood flowing in opposite directions through the petiole were to pass through alternately, rather than simultaneously. (There is no room for the blood to flow in separate channels.) In order to evaluate this possibility for the regulation of heat loss, it is necessary to study the functioning of the circulatory system both while heat is being conserved in the thorax and while it is being dumped into the abdomen.

It has so far not been possible to measure directly, millisecond by millisecond, the volumes of blood pumped in opposite directions through the petiole in an animal as small as an insect. The next best thing is to make detailed observations of body temperature changes that might be associated with heat transfer by blood, and to measure the mechanical activity of the pulsatile organs. One can then make estimates about the fluid flow and associated heat transfer.

The mechanical activities can be measured with a polygraph—an instrument commonly employed as a lie detector. It converts the organ movements into electrical signals and thence into the mechanical movement of a pen that leaves an ink line on moving chart paper—a record that can sometimes be decoded by the experimenter.

It was necessary to insert numerous probes into a bee in order to record mechanical activities of the heart and diaphragm, abdominal pumping, and thoracic and abdominal temperature. Of course bees trailing up to six wire leads could not be expected to shiver, to incubate brood, or to fly normally; nor would the probes stay inserted for long durations if the bees were active. A strong inducement was needed to encourage the tethered animals to activate the heat-transfer mechanism. That inducement was self-preservation: a thin beam of light from an incandescent microscope lamp was focused onto the thorax until high, and sometimes nearly lethal, temperatures were approached.

The tethered bees transferred externally-applied heat from thorax to abdomen—or from abdomen to thorax if the abdomen was heated. When sufficient heat was focused onto the thorax to elevate thoracic temperature, the entire abdomen increased in temperature and the rate of heat transfer to the abdomen increased greatly as the thoracic temperature began to approach lethal levels (Fig. 6.4). At a given input of thermal energy to the thorax, the bees generated a greater temperature excess in the abdomen at a high than at a low air temperature. Animals killed in place (by injecting ether into the thorax) and heated as before had only minor increases in abdominal temperature when heated on the thorax, or thoracic temperature when heated on the abdomen, indicating that the transfer of heat from the heated to the unheated body part in live animals was accomplished by some active physiological mechanism (Heinrich, 1977a).

Regardless of the difference between thoracic temperature and abdominal temperature, in live animals all internal body temperatures became nearly identical soon after thoracic heating was stopped. That is, the overheated thorax initially cooled rapidly (while abdominal temperature rose), and then all parts of the body cooled at the same rate. In dead animals, on the other hand, the cooling of the thorax and that of the abdomen were often independent of each other. All indications were that heat was being transferred within the bodies of live animals.

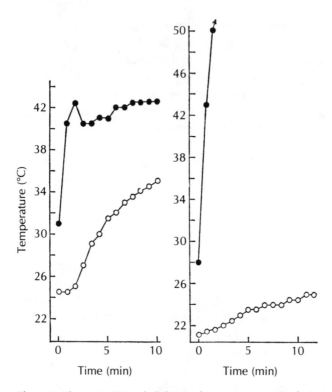

Fig. 6.4 Thoracic (●) and abdominal temperatures (○) during two heating experiments on a tethered bee. Heat was applied only to the thorax (using a narrow beam of light from an incandescent lamp). The bee prevented thoracic temperature from exceeding 42°C by shunting heat to the abdomen (left). When the heart was made inoperative, the same input of heat to the thorax caused the thoracic temperature to exceed 50°C within three minutes (right). This killed the bee because, with its tied-off heart, it could no longer prevent overheating. Air temperature was 21–22°C. (From Heinrich, 1976e.)

The mechanism of heat transfer clearly involved the circulatory system. When the pumping action of the heart was obliterated (by tying the heart with a human hair), heat transfer was abolished (Fig. 6.4). Live animals with a tied heart heated up as rapidly as dead bees, eventually dying themselves, and neither dumped heat into the abdomen.

How the circulatory system functioned to transport heat was not apparent from the experiments just described. But an examination of the interrelationships between temperature changes and the mechanical activities of the pulsatile organs gave important clues. For example, simultaneous with each diaphragm beat, abdominal temperature near the petiole rose sharply, up to 0.2°C above previous levels, and then declined to near, but slightly above, the previous abdominal temperature, in bees that had an elevated thoracic temperature (Fig. 6.5). Beats occurring in rapid succession caused a stepwise rise in abdominal temperature. It could be concluded that the transfer of heat to the abdomen was thus correlated, beat for beat, with diaphragm activity. Transfer of heat was conditional, however, on the pumping action of the heart, which generated the pressure to push the blood through the thorax. A relatively large difference between thoracic and abdominal temperatures was also required before heat transfer was apparent.

The pumping actions of the heart and the diaphragm were markedly different at high and at low thoracic temperatures. At thoracic temperatures below 40°C, the heartbeat was generally regular, very rapid, and very shallow—the heart beat weakly, but at rates of about ten beats per second. Blood was presumably entering the thorax in a thin, relatively continuous stream, thus permitting countercurrent heat exchange with blood leaving the thorax. Diaphragm beats were at almost

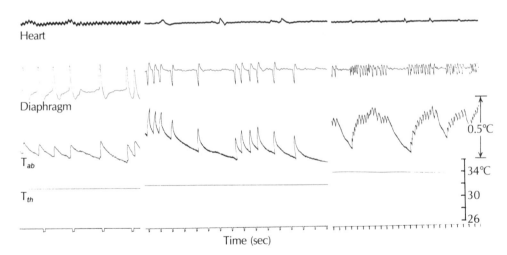

Fig. 6.5 Temperature pulses in the ventral portion of the first segment of the abdomen (T_{ab}), thoracic temperature (T_{th}), heartbeat, and mechanical activity of the diaphragm, recorded concurrently. Each beat of the diaphragm is followed by a sharp temperature pulse in the venter of the abdomen. Successive diaphragm beats result in a stepwise increase in abdominal temperature. (From Heinrich, 1976e.)

any frequency up to six beats per second, and sometimes they were absent for several seconds. The abdominal pumping—on in-out tele-scoping of the abdomen—was very shallow or absent.

A dramatic change occurred when thoracic temperature was made to approach lethal levels, around 44°C. Heartbeats increased dramati-cally in amplitude, but were halved in frequency. Diaphragm activity also became highly regular, assuming the same beating frequency as the heart (Fig. 6.6). Abdominal pumping became deep and regular, and also assumed the same frequency as the heart—about 350 beats per minute. Meanwhile, abdominal temperature, measured in the venter near the diaphragm, began to increase in a stepwise manner, each temperature step associated with a beat of the diaphragm. Tho-racic temperature stabilized or declined, if the heat applied to the tho-rax was not excessive.

How do the movements of the heart and ventral diaphragm relate to blood flow? When the diaphragm in the petiole contracts, it is raised, producing a channel beneath it through which blood can pass into the abdomen. Simultaneously, the raised diaphragm partially occludes the tracheal air tubes connecting the abdominal air sacs with the thorax. If hot and cool blood pass through the petiole alternately, then counter-current heat exchange is reduced or eliminated.

Observations of abdominal pumping movements relative to heart and diaphragm activity gave further insights into blood-flow me-chanics. In overheated bees, abdominal pumping movements were at the same frequency as heart and diaphragm pulsations (Fig. 6.7). Since resting bees that are being heated have no need for large increases in gas exchange, we can infer that the large abdominal pumping move-ments—which occurred during thoracic heating with a lamp—aided in cooling, possibly by augmenting blood flow.

Although abdominal pumping movements in bumblebees probably augment heat transfer by increasing blood flow, there is little reason to suspect, on theoretical grounds at least, that the simultaneous air movement has a large effect on heat transfer. The volumes of blood that need to be pumped to effect significant heat transfer are not unrea-sonably large, but the volumes of air would be enormous. The reason for the difference is that the heat capacity of air is about 4,100 times less than that of water, or blood, so that 4,100 times as much air as blood would have to be pumped from thorax to abdomen to accom-plish the same amount of heat transfer.

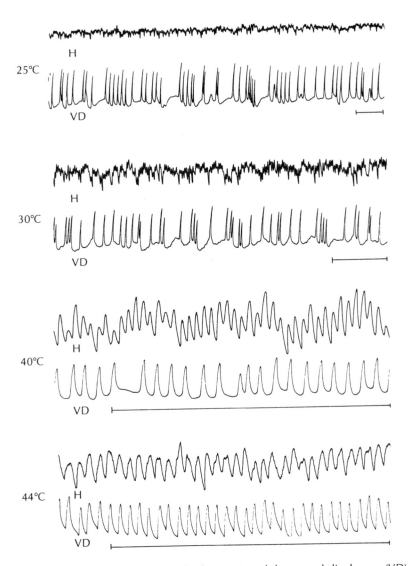

Fig. 6.6 Concurrent activity of the heart (*H*) and the ventral diaphragm (*VD*), recorded at four different thoracic temperatures (indicated at left) during heating of the thorax in a bee. The heart beats very rapidly with low amplitude at low thoracic temperatures and assumes a slower high-amplitude beat, identical with *VD* beats, at high thoracic temperatures. Five-second intervals are indicated by horizontal lines. The electrodes and amplification were not changed during the experiment. (From Heinrich, 1976e.)

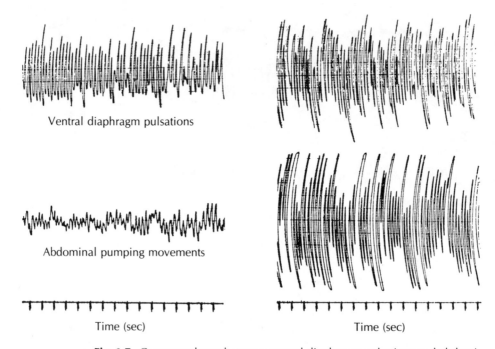

Ventral diaphragm pulsations

Abdominal pumping movements

Time (sec)   Time (sec)

Fig. 6.7 Correspondence between ventral diaphragm pulsations and abdominal pumping movements at low (24°C) thoracic temperature (left) and high (~45°C, right). The abdominal pumping movements at a thoracic temperature of 24°C correspond to a maximum of 0.5 mm displacement, while those at 45°C varied from approximately 2 to 4 mm (determined by holding a ruler underneath the abdomen). (From Heinrich, 1976e.)

Estimates of the actual volumes of air or fluid necessary to account for given amounts of heat transfer can be calculated. Let us take the example of a bee that was heated on the thorax until its abdominal temperature rose by 11°C (heated when dead, the bee increased its abdominal temperature only 2°C). From the cooling rate and weight of the dead bee's abdomen, it could be calculated that 0.53 calories of heat were actively transported per minute into the abdomen. The minimum amount of blood flow that could have accomplished this heat transfer (if thoracic temperature was within 4°C of abdominal temperature, as it is during incubation) was 0.13 ml/min. Since the heart was beating at 300 beats/min, each blood pulse needed to contain only 0.004 ml of blood to accomplish the observed heat transfer. However, using air as a vehicle for the same amount of heat transfer would have required 8 ml/second—clearly an impossibly large volume for a bee with a total body volume of about 0.5 ml. Nevertheless, evaporative water loss associated with the air movement can cause significant thoracic cooling.

What are the mechanics of blood flow in relation to abdominal pumping movements? The diaphragm in the petiole is lifted at the moment when the abdomen expands. The expanding abdomen then aspirates blood from the thorax, as well as air through the abdominal spiracles, into the abdomen (Fig. 6.8). When the diaphragm is lowered, the passage for air into the thorax is widened, and the contracting abdomen then pushes air into the thorax. The pressure from the contracting abdomen may also aid to pump blood into the thorax, particularly if the heart is also contracting at that same instant.

We do not yet know where the blood goes after entering the aortic loop in the flight muscles. It is possible that the fluid, under pressure, passes through the aorta and exits into the head before percolating back through the thorax. It may also leak out of the aorta directly into the thorax, much as our own blood pressure forces the liquid portion of the blood out of the capillaries into the kidney. Heat transfer would occur under both conditions.

In summary, the anterior portion of the ventral diaphragm probably operates like an alternating switch, first allowing the passage of both air and blood through the petiole into the thorax, and then allowing passage of blood into the abdomen. Thus, while in vertebrate animals countercurrent heat exchange is bypassed—to effect heat loss—by shunting the blood through alternate channels, in the bumblebee it is bypassed by alternating the flow of blood through the same channel. Bypassing countercurrent heat exchange makes it possible for bumblebees to dissipate heat from the overheating thorax during flight at high air temperatures, and it allows them to heat the larvae economically, without heating the entire nest. On the other hand, the bees' ability to retard heat flow to the abdomen—through countercurrent exchange—while foraging at low air temperatures means that calories are saved that would otherwise be needed for shivering to keep the flight musculature warm. Obviously the ability both to facilitate and to retard heat transfer between thorax and abdomen is of central importance to the brood care and to the energy economy of the bumblebee.

The bees have a back-up mechanism to rid themselves of heat under extreme circumstances. During flight at high air temperatures, they re-

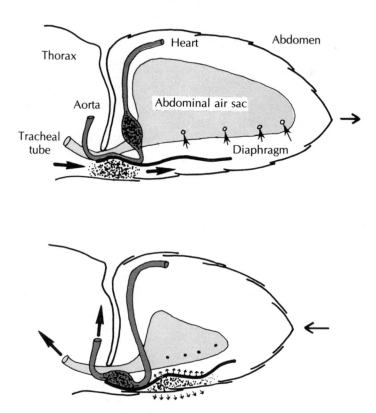

Fig. 6.8 This highly schematic diagram shows the probable sequence of heart and diaphragm pulsations—relative to abdominal pumping movements—whereby the countercurrent heat exchanger is reduced. The abdomen expands (top) and air is drawn into the abdominal air sacs from outside the animal through the spiracles (four small circles). At the same time, the diaphragm is raised and the suction created by the expanding abdomen also allows blood to enter from the thorax. When the abdomen contracts (bottom), the abdominal air sacs deflate by forcing air into the thorax. At this time, the diaphragm is lowered in the petiole, and in this position it simultaneously enlarges the air passage into the thorax while reducing the passage for blood out of the thorax. However, blood can enter the thorax by way of the heart. Small arrows indicate passage of heat from the warm blood that has entered the abdomen into the cool blood entering the thorax. In this manner, the bee is able to shunt heat to the abdomen, either to prevent overheating of the thorax or to incubate the brood. (From Heinrich, 1976e.)

gurgitate fluid from their honeycrop onto their tongue, which they "wag" in the breeze so that the fluid evaporates. They can thereby reduce head temperature by at least 2–3°C.

The bumblebees' heat transfer mechanism is not unique among insects. As already mentioned, sphinx moths also prevent overheating of their flight motor at high air temperature by controlling their blood circulation so as to dump heat into the abdomen, which acts as a "radiator" to dissipate the heat. Recently Timothy M. Casey and I (1978) found that cruising dragonflies (as opposed to "perchers") also have a fluid-cooled, rather than air-cooled, "engine." When their abdomens are waxed to prevent abdominal pumping, or when their abdominal tracheal air sacs are removed, these insects still transfer heat into the abdomen almost as effectively as intact animals. But when the heart is pinched or tied shut anywhere along the long cylindrical abdomen, including the very tip, heat transfer is immediately and totally abolished. Bumblebees and sphinx moths, as well as dragonflies, stop all heat transfer to the abdomen while they are warming up by shivering, as would be expected if they are to conserve energy and time.

Nothing is known about possible heat transfer in bees other than bumblebees. However, bees vary greatly in size and they have diverse activity patterns, and hence they have different thermal problems. They also exhibit a great variety of circulatory anatomy (Wille, 1958), which may be related to these thermal problems. Honeybees, for example, have a highly convoluted aorta in the petiole. The greatly increased surface area of the vessel in this area should ensure countercurrent heat exchange, as the loop of the heart under the abdominal air sac does in bumblebees. However, honeybees are relatively small and should have much less occasion to overheat in flight, and they do not incubate their broods with their abdomen. Thus, they have little occasion to transfer heat from thorax to abdomen. It is possible that the convoluted aorta enhances heat retention, but at the expense of the ability to dissipate heat to the abdomen.

CHAPTER SEVEN

Go to the ant, thou sluggard,
Consider her ways and be wise.
—Proverbs 6:6

Juggling Costs and Benefits

The workers of a bumblebee colony are more or less continually at
work seeking food resources and transporting them back to the colony.
Over more than 80 million years natural selection has preserved those
morphological, physiological, and behavioral traits that promote a
large net input of resources to the bumblebee factory and a large output of product. The bumblebee colony bases its entire economy on a
few components—it runs on sugar and pollen. But the management of
this economy can be complex, involving costs and benefits through
various options open to the whole colony as well as to its individual
members. At the colony level, the pollen and sugar are the resources
used to produce the machinery—combs and new workers—that will
in turn use similar resources to produce drones and new queens, the
factory's product. The colony cannot merely break even economically; it must expand as rapidly as possible, until it crashes at the end
of its cycle when it no longer invests to replace worn-out machinery,
but uses all resources to make products. Profit is ultimately measured
in terms of the number of new queens and drones that are produced
before the crash.

A worker must, over its life, not only bring in as many resources as it
expends, it must, in addition: (1) repay the colony the cost of having
produced it, and (2) make an income that can be ultimately expressed
in terms of production of sexuals.

A colony has several options for investing and utilizing its resources
to maximize output. One decision facing the colony is when in its

cycle to switch from the production of workers to the production of sexuals. The colony must also decide what size its workers will be, and hence what tasks they will be suited to perform. The workers themselves have options in the allocation of their own energy; they must balance foraging expenditures against the potential profits available from different flowers under different conditions. The various decisions and options involve immediate and seasonal, as well as evolutionary, time scales.

Machinery has to be produced before a product can be made on an assembly line, such as in a social-insect colony. There are the options, however, of beginning to make the product as soon as some machinery is available, or of investing additional energies and resources in the production of all of the machinery before finally gearing up the assembly line to turn out the product. The first strategy is inefficient, but it can yield immediate, though small, returns. The second yields no immediate returns, though ultimately the returns are potentially enormous.

Which option is more productive in the long haul? There is no absolute answer. Different kinds of bees have different strategies, each being presumably suitable to specific conditions. The numerous solitary bee species make no investments in machinery. Thus, they are able to generate a product (provision a cell with food and an egg that will grow into another reproducing individual) within several hours. This may be the only strategy possible when resources are available for only a very short time, as in deserts following isolated rains. Bumblebees have considerably more time available. They typically have one summer in which to build their factory and operate it to manufacture their product—new queens and drones.

Trying to make machinery and products simultaneously is a feasible strategy in a very capricious environment that either may or may not allow sufficient time to cash in on investments; making products without delay at least guarantees *some* output. But if a sufficiently long and predictable time-span is likely to be available for operations, then "planning" is possible. The factory can then be built to the maximum size allowable within the time constraints, and the products can then be made near the end of the available time, to utilize the factory's full potential for production while resources are still available. To some extent, the latter is the option followed by bumblebees, as well as by other annual social-insect colonies, such as those of wasps (Macevicz and Oster, 1976). This is also the option followed by many annual plants, which do not bloom and produce seed until after producing the roots, stems, and leaves needed to support their reproductive effort.

Other social bees, including the honeybees and stingless bees, have, because of long, favorable seasons of resource availability, coupled with large stockpiling capabilities, been able to escape the boom-and-bust cycles typical of temperate bumblebees and wasps. Rather than using their machinery to its fullest to make as many reproductives as possible, they export some of this machinery by swarming. The bees that swarm from the main colony help the new queens to found new colonies.

The colony also has the option of making workers—the machinery of production—of different sizes. The larger the workers the easier it is for them to regulate their body temperature, and the better able they are to forage at low temperatures. Some of the smallest workers never forage at all and spend their entire lives doing hive duties. As discussed in more detail later (Chapter 10), body size also affects the kinds of flowers utilized, because it is directly related to tongue length. Larger bees, with their longer tongues, are able to reach the nectar in long corolla flowers more easily than their short-tongued colony-mates. The short-tongued bees, on the other hand, can work faster on flowers having dense aggregations of florets with shallow corolla tubes. Thus, the colony as a whole has options that affect energy balance. But ultimately the colony energy balance is a function of how well its individual members juggle the costs and benefits of foraging.

In order to measure the bees' energy budgets during foraging, I had to find a means to measure their rates of energy expenditure in the field. The usual method of measuring the rate of energy expenditure is to measure oxygen consumption, but that method obviously could not be employed on a free-flying bee. However, a good index to a large bee's energy expenditure could be obtained by measuring its temperature as it perched on a flower and by timing the durations of its flights between flowers. The rate of oxygen consumption during flight, which had been obtained previously in the laboratory, could be multiplied by flight durations to provide an estimate of the energy expenditure of a flying bee. The energy expenditure of a perching bee, which can step up its rate of energy expenditure by exercising the muscles in the tho-

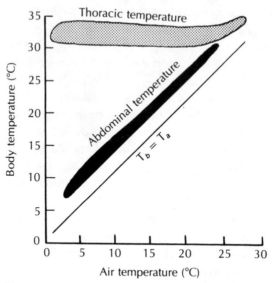

Fig. 7.1 Body temperatures of bumblebees as a function of air temperature while foraging from relatively well-rewarding flowers, such as those of jewelweed, fireweed, and milkweed. Note the large difference between thoracic and air temperature at low air temperatures. (Adapted from Heinrich, 1972a, 1972b.)

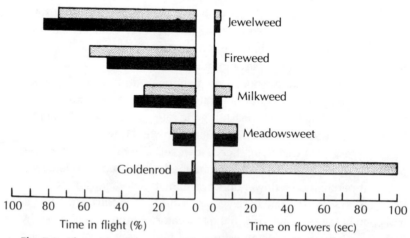

Fig. 7.2 The proportions of time foraging bees spend in flight and perching, respectively, are more closely related to the kinds of flowers visited than to the air temperature. Top bars (stippled) show both activities at an air temperature of 20°C in shade; bottom bars (solid), both at 30°C. (From Heinrich, 1973.)

rax, could be calculated from its body temperature, weight, and passive cooling rate (when dead). In order to obtain the body temperature of a perching bee I would grasp it with a gloved hand between thumb and forefinger and quickly thrust a tiny thermistor into its thorax (or abdomen), thus obtaining the temperature before the body could cool appreciably. Most foraging bees maintained their thoracic temperatures above 30°C, the minimum for flight, at air temperatures of 2° to 25°C in sunshine and shade. The body temperatures were independent of the amount of time the bees spent in flight (Fig. 7.1).

The bees were not being heated simply as a by-product of flight metabolism (see Fig. 7.2). The thoracic temperature of a perching bee can be elevated as a result of heat production during flight, provided perching durations are brief and passive cooling rates are slow, but a simple experiment showed that the bees were producing heat to keep warm while stopping at flowers. While foraging from fireweed (*Epilobium*), the bees only stopped for 1–2 seconds at each flower; the tiny drops of dilute nectar were removed with a quick jab of the tongue. However, bees would remain for up to two minutes at flowers fortified with large amounts of highly concentrated viscous syrup. In two minutes dead bees cool passively from flight temperature (32°C) almost to air temperature (12°C). But live bees that had been lapping syrup for two minutes, showed thoracic temperatures 2°C higher than bees that had continued to fly from flower to flower. Regardless of the length of time they perched on flowers, bees maintained an appreciably elevated thoracic temperature so long as they had sufficient nectar. Thus, they kept their muscle temperature sufficiently elevated for flight. Shivering is required during prolonged perching durations to counteract heat loss. The measurements showed, however, that the bees generally did not waste energy heating the abdomen—the part of the body that is not involved in powering flight.

How much energy must a bumblebee expend to maintain an elevated temperature while it is not flying? On the basis of its passive cooling rate, I calculated that in order for a medium-sized bumblebee to keep its thoracic temperature at 30°C at an air temperature of 5°C in still air in shade, it must expend a little more than half a calorie per minute (Fig. 7.3). This rate of metabolism is close to that required for flight itself. At high air temperatures the need for heat generation by the thoracic muscles is, or course, reduced. At air temperatures higher than 25°C, nearly the total energy cost of foraging comes down to what

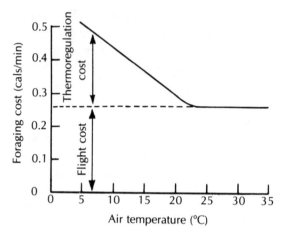

Fig. 7.3 Calculated foraging costs at different air temperatures for a bumblebee worker weighing 0.2 g that regulates thoracic temperature at 30°C and spends half its time in flight and half perched on flowers.

is needed for flight, with heat as an incidental by-product. My measurements of oxygen consumption in the laboratory have shown that the energy cost of flight is the same at all air temperatures.

Why do bees spend all that energy to regulate their thoracic temperature while they are not in flight? The answer seems obvious: they keep their flight engine revved up so that they are at all times ready to fly to the next flower. The tiny amount of sugar or pollen available in a flower can usually be taken in a fraction of a second, and since a bee cannot see the nectar inside a flower it is probing, it has no way of knowing how long it will take to suck a flower dry after its tongue contacts the nectar. However, its evolutionary experience has "told" it that it usually takes only seconds to remove all of the nectar, and the best strategy is to invest energy in staying ready for instant take-off.

A bumblebee generally visits ten to twenty flowers per minute. There must have been strong selective pressure to increase the speed of foraging because of the need to maintain a high rate of food input to the nest. The parasitic bumblebees, *Psithyrus*, which do not forage for their larvae, have a much slower foraging rate. For example, I observed these bumblebees stopping for 58 seconds, on the average, at each inflorescence of leatherleaf they visited, while *Bombus ternarius* queens at the same time and place required only four seconds to visit the 4–8 flowers of each inflorescence. While foraging, the parasitic bumble-bees maintained average thoracic temperatures 3.3°C lower than the social *Bombus*.

Armed with a background of theory and information and a thermistor and stopwatch, one can begin to investigate the energy cost of the bees' foraging in more detail and compare it with the energy supplies available. In this way one can determine if or how the bees "balance" their energy "budget" (Heinrich, 1972c). Energy income can be obtained by using capillary tubes to collect samples of nectar from flowers of given species, measuring the sugar concentration in the nectar with a refractometer (normally used by brewers to measure sugar content in beer), and then computing the energy this represents in terms of calories. For example, sugar yields about four calories per mg, and if we withdraw 1.0 μl (= 1.0 mg) of nectar that has 20 percent sugar, then this represents 0.8 calories (1.0 × 0.2 × 4.0 = 0.8). Such measurements have shown, for example, that an average fireweed blossom filled with nectar (having been shielded with screening from potential foragers for 24 hours) contains enough sugar (5.4 μl nectar with 33 percent sugar = 1.8 mg sugar = 7.2 calories) to support the maximum metabolic rate (the rate required for continuous flight or for the maximum rate of shivering when the bee is stationary) of a worker bee for about 14 minutes (7.2 cal × 1 min/0.5 cal = 14.4 min). Ordinarily, fireweed and other flowers that are good nectar producers are visited by many bees. A single blossom may be visited by several different bees within a minute, and blossoms recently visited by other foragers are generally the only ones available to bees. Most of the fireweed blossoms I examined were nearly empty, so the average amount of nectar-sugar obtained per blossom was only sufficient to support a bee's energy expenditure for a minute or less of flight or continuous shivering. In order to break even energetically, a bee would thus have to visit at least one flower per minute. Since most bees visited 20–30 flowers per minute, they were still making an energy profit while foraging from fireweed in competition with other bees.

Since different kinds of flowers provide different amounts of nectar, choosing the wrong ones can result in negative energy balance. For example, the nectar from one rhododendron blossom is equivalent to that from eleven lambkill blossoms. In order to obtain the same amount of energy, the bee has to visit eleven of the latter or one of the former (Fig. 7.4). By calculating the number of flowers of a given kind

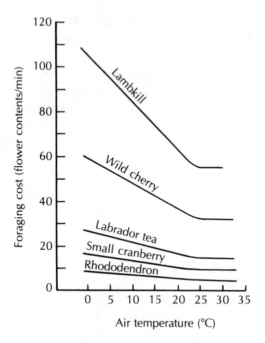

Fig. 7.4 Calculated foraging cost, in terms of the number of flowers that queen bees (weighing 0.5 g) have to visit per minute to realize a positive energy balance. It is assumed, for these calculations, that the bees spend half their time in flight and regulate their thoracic temperature at 35°C. The calculations, based on nectar available in flowers unshielded from foragers, indicate that foraging at some flowers, particularly at low temperature, is not economically feasible unless the bees make physiological and behavioral adjustments. Lambkill, although visited by queens, was exploited primarily by workers. (From Heinrich, 1975c.)

that must be emptied per unit time to cover the costs of foraging, and comparing this with the actual rates of flower visitation, one can determine how close to the energy brink the bees are operating at different flowers (Heinrich, 1975c). Since the maximum rates of flower visitation in the field can be observed, one can also calculate ball-park estimates of which flowers are energetically most rewarding, and how the balance between profit and loss changes with temperature. These calculations show that bees should be able to make a profit while foraging from rhododendron blossoms at both low and high air temperatures; they can easily visit the 5–10 flowers per minute needed to meet foraging costs and go on to accumulate profit. And indeed, the bees forage from these and other highly rewarding flowers over a wide range of air temperatures. But they cannot visit lambkill and wild cherry blossoms rapidly enough to break even energetically (which would require one-hundred and sixty flowers per minute, respectively) while foraging from them at low air temperatures (5°C). Bumblebees only forage from low-energy flowers at high air temperatures, when they do not need to divert energy for temperature regulation.

These examples only sketch a general picture of energy balance. The energy balance sheet in nature is much more complicated, and the problems of foraging optimization confronting the bee in the field are enormous. Distance from the nest, for example (which is discussed in the next chapter), is a primary consideration. Some flowers provide such minute amounts of food that they cannot be profitably harvested by bumblebees at any temperature, even when they are close to the nest. Other flowers can provide adequate net rewards only if the distance between them is sufficiently small that many can be visited per unit time, or when temperatures are not too low and competitors for the same food rewards not too numerous.

Before examining the bees' *total* energy budget during foraging we have to redefine profit from the *colony* perspective. In the narrow sense, profit can represent merely the net energy gain per foraging trip. But the same amount of food may be brought back to the colony after five minutes or three hours, for example. Aside from making a net profit per foraging trip, it is also necessary for the *rate* of profit to remain high. Time is of critical importance in the bees' foraging behavior, and the *rate* of accumulation of profit is critical for foraging strategy. "Profit" for the bees is thus best defined by: (total food intake − energy expenditure)/time. The bees can increase profit by visiting flowers at a rapid rate. This usually increases total food intake as well as energy expenditure, but profits increase nevertheless, because food intake at most flowers generally increases more rapidly than energy expenditure, which is physiologically limited.

Thoracic temperature regulation is decisive in the above equation, for it affects, indeed almost controls, the foraging rate. For example, bees with a thoracic temperature near 36°C can visit about twenty blueberry blossoms per minute, while those with a thoracic temperature near 30°C work only half as fast. With thoracic temperatures lower than 30°C bees are unable to fly. By physiologically stabilizing thoracic temperature bumblebees are able to enhance foraging intake by minimizing foraging time over a wide range of air temperatures.

Minimizing foraging time by maintaining a high thoracic temperature is not without energy costs. As already noted, the increased rewards available per unit time have to be balanced against the increased energy expenditure of keeping warm. These costs are usually low relative to the potential rewards. As we shall see, however, the bees may decrease their thoracic temperature, and their rate of flower visitation, under those special circumstances when the potential food rewards are minute and the energy cost required for minimizing foraging time results in energy debt.

In general in order to maintain an elevated thoracic temperature adequate for flight, a bumblebee worker (weighing about 0.2 g) must obtain enough sugar (about 0.14 mg) to provide 0.54 calories per minute throughout the duration of its foraging trip, regardless of the amount of time it spends in flight. The amount of energy that it must allocate for shivering declines linearly with increasing air temperature. At an air temperature of 25°C, a bee no longer needs to expend energy for warming up; if it spends 50 percent of its time in flight from blossom to blossom, as it does when foraging from dispersed fireweed blossoms, it requires an intake of only 0.27 calories per minute to break even, and if it spends 10 percent of its time in flight, as it does while foraging from flowers that are widely dispersed or require long handling times, it can make an energy profit at any intake above 0.05 calories per minute (see Fig. 7.3). Different minimum rates of flower-visitation are required before the bees can begin to show a profit at different plant species. Turtlehead blossoms, for example, are visited at rates of only two to three per minute because the bees can only slowly squeeze through the partially closed corolla of some of the flowers and because the flowers may be widely dispersed. But one turtlehead flower contains a large amount of nectar, and it can support the bees' metabolic rate for many minutes. On the other hand, hawkweed florets can be visited at up to two per second by short-tongued *Bombus ternarius* bees, because they are arranged in inflorescences that generally bloom in dense patches. But the nectar content and energy value of one hawkweed floret is minute—it barely matches the energy the bee expends in collecting it. At low temperatures bees foraging from hawkweed would expend energy faster than they could take it in, unless they reduced their energy investment by either dropping body temperature or by spending less time in flight. Potentially they could do both by remaining on an inflorescence and walking while visiting the many florets. But the hawkweed inflorescences are small, and spending more time at each one has little advantage, for the available food supply per inflorescence is soon exhausted. The tiny amounts of nectar available in these and many other composite inflorescences make it difficult, or nearly impossible, for any foragers except bees and other small pollinators to get satisfactory quantities; but they need only small amounts and their mouthparts are small enough to collect the rewards. For a person to get even the tiniest droplet visible to the human eye, for example, the whole inflorescence with its dozens of tiny florets must be squeezed at the base, and even then one is rarely rewarded with a sight of nectar. One wonders how the bees manage to make a living at these flowers at all.

Foraging profits at flowers can be increased in two ways: by increasing energy expenditure and visiting more flowers per unit time, as is possible by hovering flight, or by reducing the energy expenditure by walking from flower to flower (Fig. 7.5). Both strategies have certain costs and payoffs. Hoverers can go more rapidly into energetic debt than perchers, but they can also make more profit than perchers, depending on the food rewards they find per flower. Hovering is metabolically expensive, particularly in large animals, and hoverers cannot reduce their energy expenditure to permit foraging from low-reward flowers. Harvesting nectar from lambkill, for example, *B. ternarius* visited sixteen flowers per minute, a rate sufficient for workers (0.1 g) but not queens (0.5 g) to remain in positive energy balance (see Fig. 7.4). Because of their ability to hover and maneuver rapidly from flower to flower, clearwing hawkmoths visited fifty of these flowers per minute. Yet the moths were making even less profit than the bees. On jewelweed (with widely spaced flowers), on the other hand, the bumblebees *B. vagans* and *B. fervidus* visited ten flowers per minute, while the hovering ruby-throated hummingbird visited thirty-seven per minute, making more profit than the bees, provided the flowers were not already partially depleted of nectar. Whether foraging from very high or very low energy food resources, hoverers are always expending energy at near-maximum rates. Hovering is a feasible foraging strategy only at flowers with high food rewards. Bees must sometimes accommodate themselves to flowers with low food rewards, and they must also forage for pollen, which they can only collect by landing on flowers.

Fig. 7.5 A bumblebee (*B. ternarius*) on a goldenrod panicle, where it has the option of remaining warm and ready to fly to another inflorescence, or of cooling down and walking, utilizing many of the tiny florets.

On some flowers the bumblebees accommodate themselves to a meager supply of nectar by practicing drastic economics in the use of energy, not only perching, but remaining for a long time and minimizing energy expenditure by allowing body temperature to drop (Fig. 7.6). Bees foraging from large dense panicles of goldenrod or meadowsweet can crawl from flower to flower, using significant amounts of energy only when flying to the next panicle. When forced to feed from such plants at low temperature, because of a lack of other suitable food sources, bees often did not trouble to keep themselves continuously warm. Their thoracic temperatures sometimes dropped well

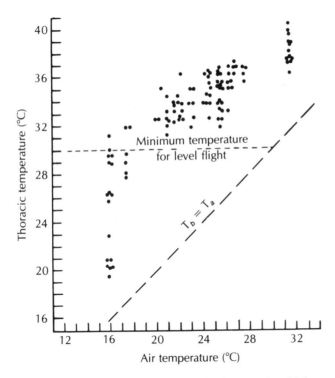

Fig. 7.6 Energy is conserved when the foraging bumblebee can walk, rather than fly, from one nectar source to the next. The graph shows the thoracic temperatures of bees (primarily drones) that were foraging on goldenrod panicles. When the air temperature was below 21°C, the thoracic temperatures of the bees frequently fell below the 30°C level required for flight, and they crawled from floret to floret. Bees that had been most recently in flight were the warmest. (Adapted from Heinrich, 1972b.)

below the threshold for flight. At air temperatures less than 20°C, many of the bees crawled over the panicles while probing for nectar and could be shaken off the blossoms onto the ground. The bees generally flew to another panicle only when the sun warmed them up or when, at infrequent intervals, they expended energy by shivering. Most of the time their engines were cut to a low idling level. I observed the same behavior in bumblebee workers foraging from densely spaced blueberry blossoms that had been depleted and contained no visible nectar. (Blueberry blossoms normally contain relatively large amounts of nectar.) In the Arctic, where bumblebees have been seen to fly at −3.6°C, they also sometimes crawl rather than fly from flower to flower.

The strategy of maintaining an energy balance by drastically dropping foraging speed is most commonly observed in drones, which only forage for their own energy needs and do not aid their sisters in collecting the food needed to complete the colony cycle before fall. Thus they are not under drastic time constraints. They often wait, before flying from one panicle to another, until heated by the sun. Use of solar radiation for heating is particularly effective with disc-shaped flowers in the Arctic that have evolved to track the sun's movements. They act as solar furnaces, focusing heat onto the pollinating insects.

Bees that are temporarily flightless through having allowed their body temperatures to drop may have as much nectar in their honeycrops as bees with a high body temperature that are foraging from more rewarding flowers. The flightless bees are clearly exerting active control by regulating their body temperature and hence their energy expenditure—they are not simply passive victims of food deprivation. The bees' energy strategy is thus finely attuned to the energy rewards available. The bees appear to be capable of a complexity of thermoregulation not observed in many "higher" organisms. Humans cannot practice energy-economy by confining heat to the muscles. And we are incapable, and intolerant, of lowering core body temperature to save precious calories, even when starving to death. However, some small mammals (for example, bats and pocket mice) and some small birds (hummingbirds) drop their core body temperature daily, although they are totally inactive when cooled. The thermoregulatory functions in the bumblebees are part of a still larger hive-centered economic strategy, where investments are not paid off until near the end of the colony cycle.

CHAPTER EIGHT

The careful insect midst his works I view,
Now from the flowers exhaust the fragrant dew,
With golden treasures load his little thighs,
And steer his distant journey through the skies.

—John Gay, *Rural Sports*

Commuting and Foraging Movements

In the field a bee spends most of its time traveling, and all the efficiency that it may derive from its flight motor is of little avail if that traveling does not achieve given objectives. When moving between food sources, bees try to keep flight time and distance to a minimum. Flight speed is of great significance to the energy economy of a bumblebee hive, since at greater speeds greater distances can be covered in less time. Nectar foragers fly at 11–20 km/hr, and they spend only 2–4 minutes inside the nest between trips. However, we do not know how far bees are willing to commute. Probably, like honeybees, they will travel at least 5 km if necessary.

From some simple arithmetic it is apparent that foraging distance makes a huge difference in the energy budget of a bee colony (Park, 1922; Ribbands,1952). It turns out that the critical cost is not the cost in energy used for commuting but the cost in *time* that could have been used to collect more food energy (Beutler, 1951). For example, let us assume that a bee that has flowers in the immediate nest vicinity can forage there continuously. Another bee foraging 3 km from the nest must spend 24 minutes in flight per trip if it flies at 15 km/h. If both bees are foraging from fireweed, which typically yields about 0.5 μl of nectar (30 percent sugar) per flower, and visiting 20 flowers per minute, then each could collect a honeycrop full of nectar (100 μl) in about 10 minutes. The commuting bee would have to fly an additional 24 minutes each time it collected a honeycrop load. Thus, its gross intake would be 30 mg sugar per 34 minutes. The bee foraging continuously near its hive could collect 102 mg sugar, or 3.4 times more, in the same amount of time. The sugar carried back to the colony by the commuting bee would be reduced another 14.5 calories, because of its additional flight metabolism over 24 minutes. This is relatively insignificant economically in comparison to the lost time, for the 14.5 calories can be retrieved from visiting 24 more flowers—about one more minute of foraging. Based on a flight-speed of 15 km/h, flowers 3 km away from the hive must be at least 3.4 times more rewarding than those close to it in order to provide the same margin of profit to the bees.

Aside from great flight speed, the bees have also evolved other mechanisms that minimize commuting time—for example, an extraordinary capacity to carry huge loads. The honeycrop is highly distensible and can occupy most of the abdominal cavity, so that the bee can carry the equivalent of 90 percent of its body weight in nectar and honey (Fig. 8.1). Pollen loads, which are carried on the outside of the legs like saddlebags, can add up to an additional 20 percent of body weight (Fig. 8.2). Thus, bumblebees (and other social bees) can carry in one trip enough food resources to pay for many trips.

The "profit" brought back to the colony on any foraging trip is the difference between the amount of sugar in the bee's honeycrop on leaving and on returning to the nest. Obviously, in order to maximize profit bees should leave the hive nearly empty, with just enough fuel to get where they want to go, plus a little reserve for emergencies. Indeed, most foragers leave the nest carrying minute amounts of honey—just enough for a few minutes of flight (Fig. 8.3). But they return carrying provisions equal to 20–100 percent of their body weight (Fig. 8.4). Some bees leave the nest with more substantial amounts, and it would be interesting to know whether these bees are foraging from distant flowers, have not yet found a source of nectar, or are foraging for pollen at flowers that have no nectar. Honeybees too take only a small amount of honey when leaving the nest to forage, and the closer their foraging destination, the less fuel they take along, thus maximizing their profits per trip.

An automated beehive developed by Tracy Allen of Berkeley has made it possible to keep precise records of bees entering and leaving a hive, as well as of the weights of their foraging loads. Allen's hive is based on an electronic tag that an untiring and nearly infallible machine (rather than a person with limited eyesight) can recognize.

Fig. 8.1 The posterior portion of the abdomen of this bumblebee has been torn away to reveal the nectar load inside the glistening honeycrop filling most of the abdominal cavity.

Fig. 8.2 The pollen is carried on the hind legs of bumblebees. The right pollen load of this bumblebee, which is collecting both nectar and pollen from mint, is clearly visible. (Photograph by E. S. Ross.)

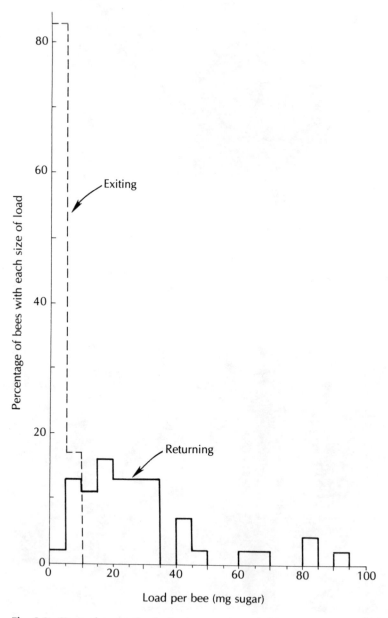

Fig. 8.3 Sizes of nectar loads (in mg sugar) carried by workers entering and leaving a *B. vosnesenskii* nest on June 7 in Berkeley. (From Allen et al., 1978.)

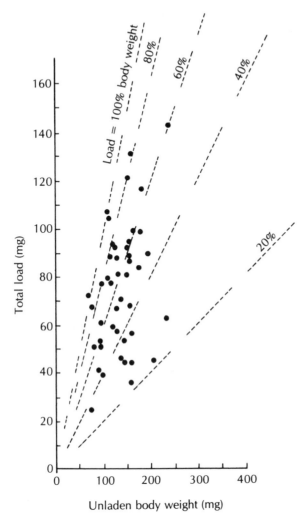

Fig. 8.4 Distribution of full loads (pollen and nectar) brought back by a random sample of forty-four workers from a *B. vosnesenskii* colony on June 7 (see Fig. 8.3). The dotted lines show total load expressed as a percentage of unloaded body weight. (Adapted from Allen et al., 1978.)

The electronic tagging system can be used to keep track of the foraging activities of individual bumblebees, to obtain detailed information on the bees' time and energy budgets, and to study the ontogeny of foraging efficiency, division of labor, and other problems of foraging behavior in relation to food resources inside and outside of the hive.

The system is based on small tags (1 mg, 2 mm diameter) that can be glued onto the thorax of each bee. Each tag contains a tiny resonant circuit that, when energized, rings like a bell with a particular resonant frequency, or tone. Each tag has a slightly different circuit with a particular resonant frequency or tone in the 500 to 1,000 megaherz (millions of cycles per second) range. A detector at the hive entrance scans the whole available frequency range, and if a bee passes under it, the tone emitted by its tag is amplified and processed for recording and computer analysis. Other pertinent information is recorded at the same time, including bee weight, measured as the bee walks over an electronic balance, and direction of bee movement, determined as the bee walks past photocells.

The tag emits a tone only when it is energized by a small transmitter, which is turned on each time a bee nears the detection area. The transmitter emits the whole range of frequencies from 500 to 1,000 megaherz. The tag, being most sensitive to its own resonant frequency, ignores all the others. Thus, the tag gives off, and the receiver picks up, only one narrow band of frequencies. Of course the whole system would be short-circuited if the receiver could hear the transmitter directly. But this is prevented by their being placed at right angles to each other.

Foraging bees seldom find enough nectar or pollen for one load at any one site. They usually have to visit hundreds of flowers over a wide area, often exceeding 500 square meters. By marking individual bees and releasing them in the field I found that experienced individuals sometimes returned to the same site each day for weeks, visiting the same clumps of plants in very similar sequences, or "trap-lines" (Fig. 8.5). After finishing a foraging trip they made a beeline back to the hive. After spending 2–4 minutes unloading at the hive, they made another beeline back to their foraging area. There was little wasted motion. Site and route fidelity minimized search and travel times for successive foraging trips. The trap-line routes are apparently learned by using visible landmarks as guides. Young bees wander about a great deal before settling down, and in uniform flat fields with no visible landmarks none of the bees established predictable trap-lines (Heinrich, 1976b).

Although individual bumblebees visited clumps of flowers in specific sequences, there was no evidence that they visited flowers within these clumps in other than a random manner. Very often they revisited

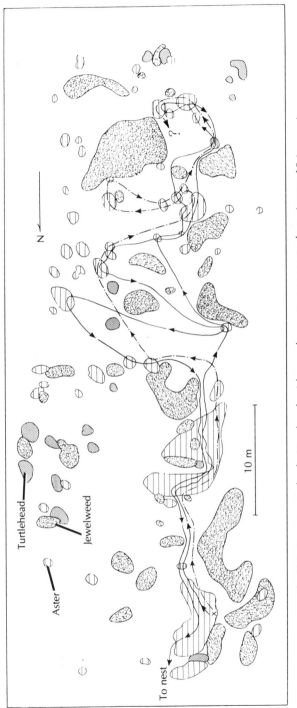

Fig. 8.5 Foraging path of a *Bombus fervidus* worker on two separate days, August 19 (———) and August 25 (—·—·—). Note that the bee's foraging area is about 50 meters in length and that the bee visited primarily aster. Many other *B. fervidus* were also specializing in aster, but most specialized in jewelweed. Each trip lasted about half an hour. The path shown here does not indicate movements within each of the flower clumps.

flowers that had just been emptied, either by themselves or by other bees in the same area. Such behavior did not appear to be adaptive. However, looking closer, we found that the revisiting of just-emptied flowers occurred generally only in patches where some of the flowers offered large food rewards. The bees could not determine what was in a flower until they had probed it, and revisiting of empties was probably an incidental by-product of the bees' efforts to locate those flowers offering large rewards. If the bees encountered only low-reward flowers in a patch, or an inflorescence, they visited a smaller percentage of the flowers and thus made fewer revisits.

The generalization that bumblebees visit flowers within clumps at random does not always hold true if the flowers are arranged nonrandomly. For example, it has been known for a long time that, when foraging from the vertical inflorescences of some plant species, bumblebees tend to fly to the base of the inflorescence and work upward. Investigating several species of nectar-feeding bumblebees foraging on the vertical inflorescences of several *Delphinium* and *Aconitum* species, Graham Pyke found that 90 percent of the time workers began foraging at the bottom of each inflorescence. The bees then tended to move vertically up the inflorescence from one flower to the next. However, they missed about one-third of the flowers and they usually left the inflorescence well before reaching the top flower. The movement of the bees minimized the revisitation of just-emptied flowers, and it may have additional significance, as Lynn S. Best and Paulette Bierzychudek (1978) have shown in a study of bumblebees foraging from foxglove.

Foxglove is a European plant with large, bell-shaped, white or purple flowers arranged in vertical inflorescences of ten or more flowers (Fig. 8.6). It is now well-established in Washington and British Columbia, where it is pollinated primarily by bumblebees. Each day a new flower opens at the top of the inflorescence, and a flower at the bottom withers. Individual flowers remain open for about ten days. Each flower is first functionally male and then female; the anthers mature first and the stigma becomes receptive later. The "male" flowers are at the top of the inflorescence and the "female" flowers are at the bottom. In "full" inflorescences (which had been screened to exclude all foragers) the older (lower) flowers always contained more nectar than the upper flowers. If bees are foraging optimally—so as to get the most nectar in the shortest time—they should begin to forage at the bottom

Fig. 8.6 An inflorescence of foxglove, *Digitalis purpurea*.

of an inflorescence, move up, without revisiting just-emptied flowers, and leave when profits become unacceptably low. In about 80 percent of the observed first visits, the bees started at the bottom third of the inflorescence. Seventy-seven percent of the moves were up, and only 3 percent of the visits were to just-emptied flowers. The bees left the inflorescence after having visited 4–5 flowers. In contrast, while visiting "empty" inflorescences (those that had not been protected from foragers), 63 percent of the bees left after having visited only one flower. These results show that the bees respond in their foraging movements to changes in nectar availability.

The pattern of nectar rewards, which elicits the characteristic bee movement on foxglove, promotes cross-pollination. The bees are encouraged (1) to visit the female flowers first, fertilizing them with pollen from other plants, and (2) to remain on the plant long enough to remove pollen from at least one male flower. These general results and conclusions are in agreement with Pyke's observation on *Delphinium* plants, which employ a pollination strategy similar to that of foxglove.

Bumblebees change their foraging behavior in other ways in response to varying nectar abundance. The more nectar they find per flower, the more they restrict their search to the surrounding area—to nearby inflorescenses, to the flower patch where the rewarding flowers are located, and to neighboring patches. In one set of experiments we laid screening over patches of white clover blossoms (or heads, each of which had 15-40 individual florets), to let nectar accumulate. Flowers outside the screened area were visited all day by bumblebees. The screening was then removed and the bees could then visit both nectar-rich and nectar-depleted areas. The behavior was markedly different in the two areas. In the depleted areas bumblebees probed on the average only two florets on each flower head, while they probed twelve florets in the nectar-rich areas. When bees entered nectar-rich areas they stayed there for long durations, but bees foraging in depleted areas moved from one part of the field to another. In the rich areas the bees made short flights and flew with sharp angular movements, rather than the long flights and broad turning movements they made between successively visited flower heads in the depleted areas (Fig. 8.7). Thus, the bees concentrated their foraging in those areas where it was most likely to be rewarded, and moved quickly through areas of low food rewards till better forage was encountered.

The bees have an additional option to face in their efforts to maximize the returns from foraging: they can stay at each flower until they have picked up every last scrap of nectar and pollen, or they can hurry from flower to flower, skimming the cream and returning for the dregs later, when the major supplies have been removed. Tom Witham (1977) has investigated how bumblebees vary their behavior in response to variations in nectar distribution in the desert willow (*Chilopsis*). Nectar is the only reward offered by the flowers of this plant to bumblebees and other large bees, its pollinators. About 1.1 μl of nectar is held by capillary action in five grooves radiating out from the base of the corolla (Fig. 8.8). When the grooves are full, an additional 8μl accumulates in a pool at the base of the grooves. *Bombus sonorus* queens can remove pooled nectar at the rate of 2.0 μl/sec, but because each of the five grooves must be individually probed, they can remove the groove nectar at only 0.3 μl/sec. Witham observed that when bees removed all of the nectar they probed first into the nectar pool and then into each of the grooves. It turns out that, in order to take in the most nectar per unit time when food is abundant, bumblebees should specialize in pool nectar and not bother with groove nectar. When nectar abundance declines, they should remove all of the nectar. This

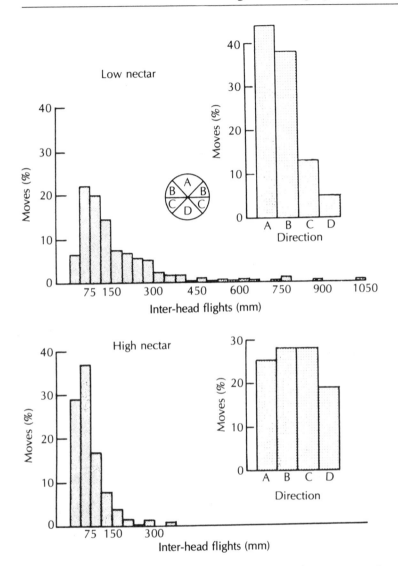

Fig. 8.7 Changes in foraging behavior of *B. terricola* workers in two patches of white clover, one that was utilized by many bumblebees and had only 0.003 mg sugar per flower (top), and one that had been screened with bridal veil to allow nectar to accumulate to a level of 0.01 mg sugar per flower (bottom). The left-hand graphs show that the bees skipped over many flower heads when nectar rewards were low. The right-hand graphs show that they no longer persisted in moving in the same direction after successively visiting flowers that contained high food rewards (the directions of movement are as shown in the clocklike figure—*A* is forward, *D* backward).

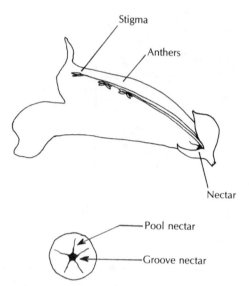

Fig. 8.8 Flower of desert willow, *Chilopsis,* with cross-section near base of corolla tube showing pool and groove nectar.

is precisely what the bees did. Witham measured the amount of nectar in *Chilopsis* flowers at different times of the day, as well as the amount of nectar removed from a flower in a single visit. Early in the morning the flowers had the largest amount of nectar (2.4 μl per flower), and by 9:30 A.M. nectar volume had declined to 0.3 μl per flower as a result of the bees' foraging. Early in the morning the bees specialized on pool nectar, reducing the nectar content of each flower from 2.4 to 0.7 μl (Fig. 8.9). By skipping the groove nectar they were able to move rapidly from one flower to the next. However, when total nectar per flower had declined to 0.2 μl, the bees took the groove nectar. By taking into account the time needed to suck up nectar, and the caloric cost of foraging at different temperatures, Witham calculated that in the early morning the bees that took only the pool nectar were making a net caloric gain of 12.3 calories per minute, while if they had taken all of the nectar of each flower they would have made a net gain of only 9.9 calories per minute, a 25 percent decrease in foraging efficiency.

The plant's evolution of this nectar-dispensing system allows it to be visited at least twice, first by high-energy-demanding bees and second by food-stressed bumblebees or any of the small, energetically less demanding solitary bees that visit the same flowers. In effect, the plant is increasing the number of pollinator visits per given nectar reward by partitioning the rewards.

The potential for foraging improvement is apparently inexhaustible. For example, recent observations in Central America by Gordon Frankie and his co-workers (1976) suggest that some solitary bees of the genus *Centris* have found a better solution to the problem of inci-

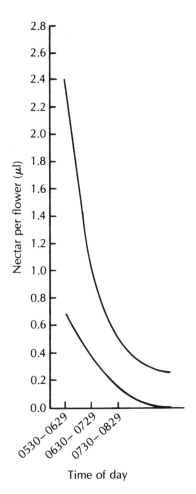

Time of day

Fig. 8.9 Amount of nectar in *Chilopsis* flowers at different times of day. The upper line represents the average standing crop in an unvisited flower, and the lower line represents the average amount of nectar remaining after a bee had visited the flower. The difference between the two is the amount of nectar removed by the bee. (From Witham, 1977.)

dental revisiting of just-visited flowers during foraging. Tight groups of 15–300 foraging bees move as "waves" from branch to branch in densely flowering trees. They chemically mark, and subsequently avoid, the just-cleared areas, thus they avoid crossing their own and each other's paths and forage only in previously unexploited areas. Curiously, the groups include only male bees; they sleep together in clumps in mid-afternoon and at night and set up individually-defended territories from which they chase all other males and where they intercept females.

In summary, time is of prime importance in the bees' race to harvest resources. Large honeycrop capacity, short foraging distance, and great flight speed all increase the time available for foraging. The costs of carrying large loads and flying long distances are insignificant, in terms of energy expenditure, when compared with the cost of food energy not collected because of lost foraging time. Foraging bees travel between, and sometimes within, clumps of flowers in patterns that act to increase foraging returns.

CHAPTER NINE

Wiser far than human seer,
Yellow-breeched philosopher!
Seeing only what is fair,
Sipping only what is sweet,
Thou dost mock at fate and care,
Leave the chaff, and take the wheat.
—Emerson, *The Humble-Bee*

Foraging Optimization by Individual Initiative

A bee starting to forage in a meadow with many different flowers faces a task not unlike that confronting an illiterate shopper pushing a cart down the aisle of a supermarket. Directly or indirectly, both try to get the most value for their money. Neither knows beforehand the precise contents of the packages on the shelf or in the meadow. But they learn by experience.

The proposition the bee faces is even tougher than that confronting the human shopper. Some of the products on nature's open market are designed to deceive. They advertise rewards but contain no goods. There are no fair trade laws and there is no consumer protection. There are sometimes temporary bargains, but those on the market today are often gone tomorrow.

Which of all the available flowers do individual bumblebees visit? And how do they manage to maintain the flow of food into the hive economy as the floral resources change through the season? It was previously thought that bumblebees were inconstant foragers—since 50 percent or more of pollen loads brought into the nest were derived from two or more (and sometimes more than six) kinds of flowers (Brian, 1950; Free, 1970), whereas no more than 2 percent of honeybee pollen loads were of two kinds (Free, 1963). Information on pollen loads was interesting as far as it went, but it did not reveal much about the foraging behavior. Pure pollen loads might merely reflect the availability of only one kind of flower at the bee's foraging area, and mixed loads could reflect random foraging at a site with many kinds of flowers.

In order to distinguish incidental from true flower constancy, I followed individual bumblebees in the fields of our farm in Maine, which had been studiously neglected for several years and contained a variety of flowering plants at any one time. I found that although bees of any one *species* were foraging from a variety of flowers (though tending to favor some over others), the *individuals* generally concentrated on a small spectrum of those available, independently of any apparent preferences of their species. For example, in a hayfield with simultaneously flowering red clover, fall dandelion, and wild carrot, all species of bumblebees were foraging from all kinds of flowers available. However, the largest percentage of B. fervidus were foraging from the fall dandelion, most of the B. terricola were on the wild carrot, and B. vagans tended to be most common on red clover. By following individuals and noting each flower they visited, I found that one B. vagans individual, for example, was visiting fall dandelion primarily and wild carrot secondarily, by-passing the clover in its path. Another B. vagans in the same area of the field was visiting wild carrot primarily and clover secondarily (Fig. 9.1). It seemed appropriate to use familiar academic jargon and say that individual bumblebees had major and minor specialties in their foraging repertories (Heinrich, 1976b).

Some bees were remarkably consistent in their foraging behavior. The same individuals (identified by numbered tags glued onto the thorax) would come back trip after trip, day after day, to the same area and to the same flowers. I observed several B. fervidus workers almost every day for a month and found that some foraged almost exclusively on jewelweed, some on aster, and some on goldenrod.

In one area where jewelweed was growing closely intermingled with aster, I removed all of the jewelweed blossoms I could locate. The area then had a nearly continuous cover of aster flowers, and the remaining jewelweed specialists flew over and among them as if oblivious to their food potential. Meanwhile, the aster specialists continued their foraging, without at any time interfering with the jewelweed bees, which at times landed briefly on the asters. Some jewelweed specialists came back and searched many days for the few isolated blossoms of their specialty that I had inadvertently failed to remove. Still other former jewelweed specialists visited almost all of the remaining kinds of flowers—asters, goldenrod, and turtlehead—and eventually adopted one of them as their new specialty.

The main advantage of specializing is that it allows the bees to de-

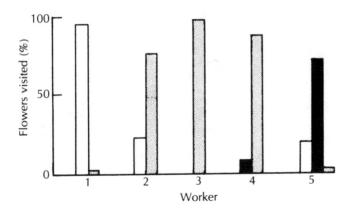

Fig. 9.1 Flower preferences of five different *B. vagans* workers simultaneously foraging in a field with three kinds of flowers intermingled. Open bars = fall dandelion. Stippled bars = red clover. Filled bars = wild carrot. (From Heinrich, 1976b.)

rive the full economic benefits that a particular flower may offer. First, bees can find proven sources of food more quickly if they can search for, and recognize, specific flower signals that can be seen from a distance. Second, the different flowers have different morphologies requiring specialized skills for efficient manipulation. For example, to collect pollen from wild carrot a bee must rapidly walk across the flat inflorescence, pressing its body down to pick up pollen from the surface of the many tiny florets. In order to collect pollen from the wild rose, a bee grasps groups of anthers, vibrates them, turns, and grasps a second group of anthers, and so forth. To collect pollen from timothy grass a bee simply scrambles up an inflorescence (Heinrich, 1976d). The pollen of this wind-pollinated plant comes loose easily. In contrast, the pollen in blue bindweed is contained in tubular anthers, and a bee must grasp the flower with its mandible and shake the pollen loose by vibrating its wings. The buzzing vibrations that shake the bee and the flower it grasps release the pollen into the venter of the bee, and from there it is transferred to the corbiculae on the hind legs.

Nectar collecting is as specialized as pollen collecting. Probing into raspberry flowers, or into the open inflorescences of asters and goldenrods, may require little skill, but to enter a turtlehead flower a bee has to pry apart the partially closed lips of the corolla and probe for nectar deep within the interior of the flower (Fig. 9.2). To successfully utilize

Fig. 9.2 A *B. vagans* worker prying open and entering a turtlehead blossom. The partially closed corolla tube excludes most other foragers.

Fig. 9.3 A bee (*B. fervidus*) entering a jewelweed flower becomes dusted with large amounts of pollen, but this pollen, which is apparently unpalatable, is discarded. The bee is cleaning its dorsal surface with its middle pair of legs.

Fig. 9.4 A short-tongued bee (*B. terricola*) robbing jewelweed of nectar by biting into the nectar spur of the flower rather than probing through the legitimate entrance. (This bee cannot reach the nectar in the long jewelweed nectar spur from the entrance.)

iris flowers a bee must land on the large wing petal and crawl upward and in. A bee can extract the nectar from jewelweed blossoms either "legitimately," by probing through the front entrance over the reproductive organs of the flower (Fig. 9.3), or "illegitimately," by biting through the tender nectar spur, thus robbing the flower without pollinating it (Fig. 9.4). Bumblebees collect only nectar from iris and jewelweed flowers, although from many other kinds they simultaneously collect both pollen and nectar.

A good example of how specialization aids foraging success, and of how unspecialized individuals may fail to reap rewards for their foraging efforts, is provided by monkshood flowers. The nectar in these morphologically complex flowers is hidden in the tips of two modified petals under a hood consisting of modified sepals. When I observed bumblebees visiting monkshood flowers for the first time, I noted that they were having difficulty, even though this flower is considered to be a classic bumblebee flower because it is pollinated exclusively by bumblebees wherever it grows wild. Many bees that had not visited these flowers before did not reach the available nectar at all, and they sometimes approached buds as well as flowers. They sometimes entered the top of a flower rather than going in at the bottom over the anthers (the usual approach). The monkshood specialists, on the other hand, moved rapidly from one flower to the next, always entering by the easiest route (over the anthers) and always reaching the nectar in one smooth motion (Fig. 9.5).

In early spring I have seen many freshly emerged bumblebee queens probing for nectar in male pussy willow catkins, which have none. Similarly, in a field with cinquefoil blossoms—flowers that provide pollen only—some bees tongued the flowers as if expecting nectar. But the specialists on cinquefoil vibrated the flowers, shaking loose the pollen without attempting to collect nectar. All gradations of behavior, from 0 percent to 100 percent foraging efficiency, were observed. There is probably no improvement in a hive with time, since new bees are born every day. Obviously, specialization on only one flower kind is the best foraging strategy for any bee at any one time—provided the bee is specializing in the most rewarding flowers. But is it advantageous for bees to have minor, as well as major, specialties? Straying from strict specialization would appear to be disadvantageous, once the best flowers have been identified.

Simple experiments showed, however, that minoring was some-

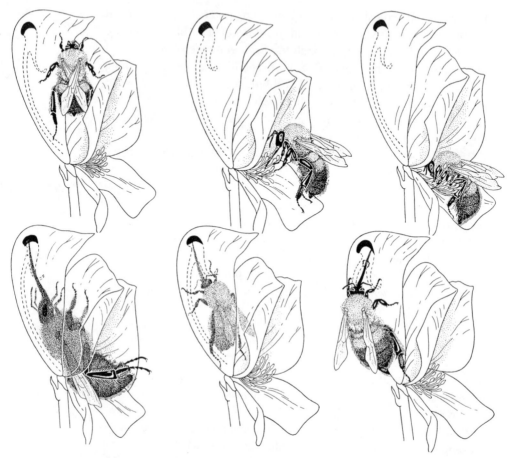

Fig. 9.5 Bumblebees manipulate monkshood flowers in different ways. In the top row an inexperienced bee attempts to enter a flower from the top (left) and probes for nectar among the anthers (center). At top right a bee is vibrating the anthers to collect pollen. The bottom row shows typical ways in which large and small bumblebees reach the nectar, and a bee robbing nectar without entering over the anthers. (Adapted from Heinrich, 1976b.)

times adaptive. When minor flowers were experimentally made more rewarding, bees made their minors their majors. For example, I used large sugar syrup rewards to enrich the minor flower—the aster—of a bee majoring in goldenrod. The bee then adopted the aster as its new major (Fig. 9.6). (But the longer the bees major, the less readily they switch.) Minoring is thus a compromise behavior that can be seen as

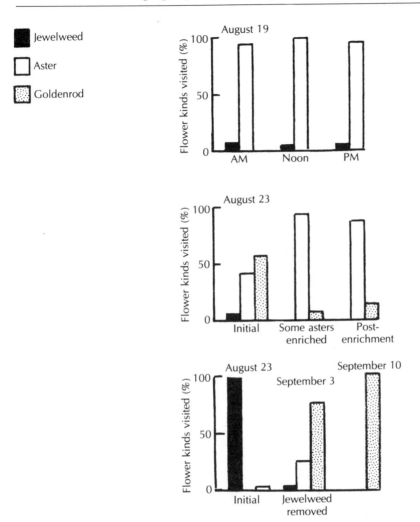

Fig. 9.6 Individual foraging preferences and changes of preference for three *B. fervidus* workers at a site where aster, jewelweed, and goldenrod flowers were in bloom. The top graph shows portions of three foraging trips of one bee at different times of the day, demonstrating its flower-constancy. The middle graph shows three consecutive and full foraging trips of another bee before, during, and after enrichment of some of the "minor" flowers with sugar, demonstrating fluctuation in preference. After the last post-enrichment trip, the bee deserted the area. Bottom: portions of three foraging trips before and after the "major" flowers (jewelweed) were experimentally removed from the foraging area; the bee resampled the flowers and then switched to goldenrod. (From Heinrich, 1976b.)

advantageous when viewed from the perspective of *changing* resource availability—it allows bees to keep track of, and to respond to, fluctuating resources or needs. The bees play a game analogous to the stock market. They do not know beforehand which is the most upcoming commodity (flower), and their best strategy is to invest primarily in the flower that appears to be the most remunerative while simultaneously investing some energy in several minor species. When the rewards of a minor go up, investments can be shifted appropriately. In summary, bees in any one time and place collectively possess a wide range of foraging skills that allow them to fully utilize all, or nearly all, of the available flowers, and to keep abreast of changing resources.

Among bees, success in competition for resources depends on making the proper flower choices and having the right foraging skills, acquired through individual learning and specialization. Flower specialization that was genetically fixed, as found in many solitary bees, would obviously not work for bumblebees and other social bees that require a continued food input for the hive economy. Social bees rear successive batches of workers throughout the growing season, and since a succession of different plants will have bloomed the colony cannot base its entire food economy on any one of them.

Is learning an important component of foraging behavior? How do bees decide to adopt certain majors and minors when they initiate their foraging careers? Does this involve random choice, communication, and a coordinated hive response, or individual initiative? In order to find out, it was necessary to observe foragers on their first and subsequent foraging trips out of the hive. This would be a nearly impossible undertaking in bees having unrestricted movement in the field. I therefore built a large screen enclosure over a portion of meadow with many different plants in bloom. Additional flowers were planted to stock this bee supermarket. Then I introduced a small colony of bumblebees in a hive-box with an entrance hole that could be plugged or unplugged with a stopcock. I let out one bee at a time, watching its movements and flower visits until it returned to the stopper at the closed entrance to be let in again. Then the procedure was repeated with another bee, and later on with the same bees on successive foraging trips. After I had spent one summer watching the bees in the enclosure and several summers watching individual bees from morning till night outside the enclosure in the field, some patterns in foraging behavior began to become evident.

The bumblebee workers sometimes initiated their foraging careers when they were only two days out of the cocoon. Regardless of age, however, all new foragers picked flowers more or less indiscriminately on their first foraging trips. They visited flowers with no nectar or pollen, or very little, as well as those with ample rewards. But sometimes they tried to collect nectar from flowers that contained only pollen, and vice versa. They were usually unable to reach the rich nectar supplies on their first tries at morphologically complex flowers with hidden nectar.

Inexperienced bees were inefficient foragers for at least two reasons. First, they foraged for nectar from low-nectar flowers like goldenrods and asters when high-nectar flowers like jewelweed and turtlehead were available. There was, however, some advantage for the inexperienced bee in foraging from those flowers, since they are morphologically simple and no skills are required to extract the nectar and pollen. Nevertheless, the bees still needed about an hour to collect a load of nectar from such flowers, while those bees that had found and learned to utilize turtlehead or jewelweed flowers could collect a load in about six minutes.

A second reason why inexperienced bees were inefficient foragers is that they did not restrict their foraging to specific routes or tracks to minimize travel time between rewarding flowers. Third, morphologically complex flowers like turtlehead or jewelweed were often handled inappropriately, so that the rich nectar contents were not harvested even when the flowers were visited. Those bees that eventually did learn to manipulate these complex flowers collected nectar loads in about 6–7 minutes. After having mastered entry, they also attempted to force their way into flowers that had not yet opened. They did not always succeed, and superficially their behavior thus resembled that of inexperienced bees. Their behavior was different, however, in that they only attempted to enter the difficult flowers at those precise places where the flowers normally opened.

After two to six foraging trips most of the bees in the enclosure became expert shoppers. They found the best bargain—jewelweed—and they learned to handle these flowers correctly to get the nectar (Fig. 9.7). At that time the experimental colony had surplus pollen stores and no larvae, so that they did not have to visit pollen flowers. They specialized on that nectar source, jewelweed, yielding the most profits in the shortest time, a honeycrop full in six minutes (Fig. 9.8). As

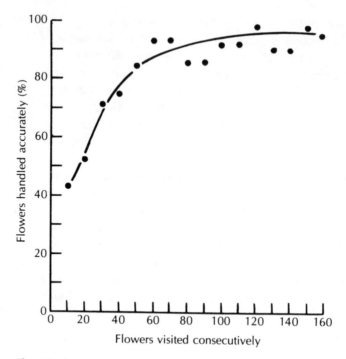

Fig. 9.7 Improvement in handling accuracy of *B. vagans* workers at jewel-weed flowers, starting with the first flower each encountered in its foraging career, and proceeding to the 150th flower. (Adapted from Heinrich, 1978.)

concluded previously by Hobbs (1962), each bee probably arrived at its choice of flower independently. If the bees had communicated one would have expected them all to start utilizing the same flowers more or less simultaneously, but different individuals started to specialize on jewelweed at different times. These times were apparently related to individual random discovery times of these flowers. Those bees that stayed with their original choices—goldenrod and aster—were the least likely to discover the more superior foods available.

It was somewhat of a surprise to note that the jewelweed specialists did not minor in other flowers. These results did not fit fully with a mathematical model predicting what the bees should have done if they were foraging optimally (Oster and Heinrich, 1976). The model, based on previous observations of bees picked at random in the field that were majoring and minoring, indicated that minoring was a necessary compromise to keep track of changing resources. Apparently, however, the bees maximize immediate profits when food rewards in their specialty exceed some level.

The human shopper may also become conditioned to prefer certain items, but unlike the bee he or she generally tries to vary the diet. In the bee's diet there are only two necessities: nectar and pollen. Different flowers vary primarily in their packaging of these two items, so bees have little to gain nutritionally by visiting different kinds of flowers. Also, a varied diet can be attained even if individuals are highly specialized in their foraging because the goods are redistributed in the nest. Individuals can specialize in the collection of either nectar or pollen, although they usually collect one product incidentally to the other from those flowers that offer both.

On the open marketplace or in the field good deals are soon exploited. Our bumblebee shopper faces a radically different situation after innumerable competitors have had equal access to the nectar and pollen available. As the best flowers are depleted, their value approaches that of the less desirable ones (Fig. 9.9). The time required to

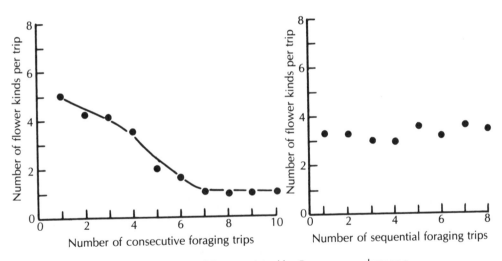

Fig. 9.8 Number of different kinds of flowers visited by *B. vagans* workers as a function of number of consecutive foraging trips (left), starting with the first flower of each bee's foraging career. Competition was held to a minimum (only one bee at a time had access to the flowers in the enclosure). When all of the colony specialists were allowed to forage simultaneously (right), they depleted the jewelweed nectar and during subsequent trips began to visit several kinds of flowers. (Adapted from Heinrich, 1978.)

gather a given quantity of sugar from them increases greatly. For example, in the enclosure bees at first required about 6 minutes to collect a nectar load from unexploited jewelweed, and 60 minutes from goldenrod. But after I let all of the jewelweed specialists forage simultaneously, foraging time on jewelweed increased to 30–40 minutes, while foraging time to collect a load of nectar from the less-utilized goldenrod changed little or not at all.

When the nectar values in jewelweed became low because of competition from hive-mates, experienced bees became as inconstant to these flowers as they had been originally to the low-nectar flowers. They again broadened their choices (Fig. 9.8). Although still tending to adhere to individual major specialties, they again visited all sorts of flowers and they retained minor specialties in their foraging repertoires. Although the flowers may have been similar in nectar content, it is possible that the bees *perceived* them to be slightly different, because the different handling skills of individual bees yielded different rates of nectar extraction on different flowers.

The bees in the enclosure, under competition, visited more kinds of flowers per foraging trip (or per given number of flowers visited) than those outside the enclosure, which were experiencing at least as much competition. This variation could have been caused by the number or size of patches. In both the enclosure and the field the bees often visited the same flower patches and the same individual flowers (though probably not intentionally) on successive foraging trips. But they did not stay indefinitely in any one patch. They had a tendency to move from one patch to another, and when they ran out of patches of a given flower kind they visited other flower kinds. In the field, however, there was usually a nearly infinite number of large patches of each flower kind, and once having established a specialty, bees have less tendency to switch as long as their favorite flowers remain available.

We decided that it would be fun to switch our study from bees in the field to bees foraging in the laboratory. It was essential that we use artificial flowers for the experiments because we wanted to isolate, or at least control, the stimuli that the bees might be responding to. Natural flowers have different scents, geometric patterns, and colors. It is difficult, if not impossible, to vary these potential stimuli independently of one another. Furthermore, natural flowers are too short-lived to be used in foraging experiments. Making simple flowers out of standard colored paper or tape was easy, but designing a system to deliver pre-

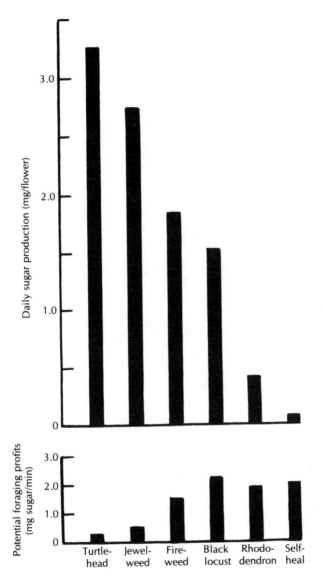

Fig. 9.9 Daily sugar production of *individual flowers* (top graph) and net sugar rewards that were available *per minute* of foraging time (based on number flowers visited per minute and amount of reward left per flower) in flowers left open to foragers (primarily bumblebees). In part because of the depletion of flowers with high rates of nectar production, many of the flowers end up offering roughly similar food rewards per given foraging effort. (From Heinrich, 1976b.)

cise amounts of nectar into the flowers, according to specific feeding regimes, proved to be much more difficult. We designed a foraging arena of about one-half a square meter in which the bumblebees appeared to forage normally. Small holes drilled into the floor served as syrup wells, and these food sources were identified by artificial flowers (stars or other patterns of different colors) placed directly over them. The bees probed through an aperture in the artificial flower into the nectar well beneath. The nectar wells were manually refilled with minute but precisely known quantities of syrup by a push-button device that delivered as little as 0.5 microliters at a time through a small polyethylene tube.

After spending an entire summer, from morning to night, sometimes seven days a week, training bees by feeding them as little as .5 μl syrup at a time, we learned that the bees had certain innate preferences that determined rates of learning, rates of forgetting, and ability to switch to more rewarding food sources (Heinrich, Mudge, and Deringis, 1977). When the bees were given a choice between two rewarding flowers of one color and two nonrewarding flowers of the other color, they eventually learned to restrict themselves only to the rewarding flower. After about two hundred visits to white flowers having 1.0 μl of 50 percent sucrose solution per flower, the bees restricted themselves to white, but with blue flowers most bees became flower-constant after visiting fewer than fifty flowers. And when *both* white and blue flowers were rewarded, but blue six times less than white, they usually visited both, although some individuals stayed with the low-reward blue (Fig. 9.10). When the bees had been trained to one color only they stayed with that even after we rewarded the oppositely colored flowers. We then rewarded the bees less and less frequently at the flowers to which they had been originally trained, simulating the flowering succession of species in the field. We found that if the bees had been trained to white they readily switched to blue. But the reverse was generally not true.

We interpret the results as indicating that the bees learn blue more readily than white. Perhaps this is because many bumblebee flowers are blue, and such flowers have proven themselves to be suitable through evolution. Moreover, caution on white flowers is advantageous, because many white flowers, pollinated by flies, contain minute amounts of food. Even when we made white flowers more rewarding than blue the bees never fully abandoned the empty blue flowers. This cautiousness in the use of white flowers, which has also

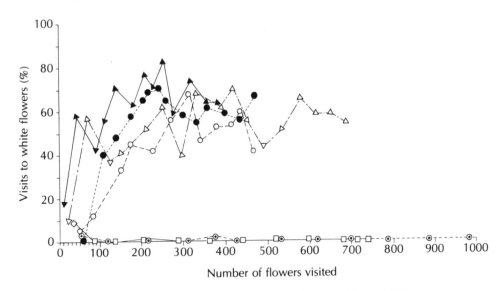

Fig. 9.10 Percentage of visits by a *B. terricola* worker to white artificial flowers, as opposed to blue flowers, when each visit to white was rewarded with 3.0 μl 50 percent sucrose, while each visit to blue was rewarded with 0.5 μl 50 percent sucrose. The different symbols depict different individuals. Note that despite the inferior food rewards in blue, two individuals visited these flowers exclusively, and the others continued to visit them occasionally. (Adapted from Heinrich, Mudge, and Deringis, 1977.)

been found in learning experiments with honeybees (see Menzel, Erber, and Masuhr, 1974), may result simply from the bees' inability to remember white as well as blue. Performance on white increased from one foraging trip to the next over 3–4 hours on the same day, but the performance declined during the following 20–21 hours overnight. The percentage of white flowers visited eventually stabilized when the increment in performance during the day equaled the decrement occurring overnight. The bees switched to a new flower color—blue, for example—when the memory for white had faded. Since the memory for blue was, in part, genetically fixed, the bees did not forget it, and they seemed relatively resistant to switching to white after having their inherent bias for blue reinforced.

Let us now briefly review the general plan of bumblebee foraging, before discussing it from still another perspective. Bumblebees lack recruitment ability, and they cannot overpower other bees and rob their nests, nor can they hold and defend feeding territories as stingless bees do, or be directed to rewarding food by hivemates, as honeybees do. As far as we know each forager must individually determine which flowers are best at any one time. At the beginning of their foraging careers individual bumblebees sample a wide variety of flowers over a wide area. At such times, they are nearly impossible to follow because they fly rapidly and do not stay long in any one place. Eventually they learn to manipulate morphologically complex flowers and to avoid nonrewarding flowers. They become site- and flower-specific, establishing foraging routes between rewarding clumps of flowers. Maximum foraging profits are possible only if the bee makes the correct choice of flowers and manipulates these flowers correctly in order to gain all of the potential food rewards. When most of the high-nectar flowers have become depleted and all flowers offer low, and roughly comparable, food rewards, then flower choice is of lesser importance. Whether or not a profit is made will then depend even more strongly on foraging skill, particularly if the flowers are morphologically complex and require precise manipulative behavior for entrance or extraction of rewards.

The bees of a colony have a division of labor, in that they divide up the foraging from the different sources that require different skills. However, the division of labor is not based on any plan. The bees simply respond to the work as it is going on. They take on the work that is available to them. Thus the flower that yields the most net profits will attract the most bees, while another flower offering lesser profits will have fewer bees utilizing it. A group work pattern results, but it is made up of the interactions of the individual patterns. Even though the bees are free agents, an order ensues out of their combined actions, as if each individual were led by an invisible hand.

Is the pattern of foraging in which individual bumblebees attempt to reap the largest possible rewards really the best for a colony? It seems possible that some workers might contribute more to the colony stores by foraging from suboptimal flowers, rather than by foraging optimally as individuals. For example, if worker *1* visits flower *A*, gaining 1.0 units of energy, and worker *2* visits flower *B*, gaining 0.4 units, then the net colony gain is 1.4 energy units. But if worker *2* is trying to maximize *individual* gain it should, like the bees restricting themselves to fortified flowers, forage from flower *A* rather than *B*. But if it also forages from the exact flowers that bee *1* is using, then bee *2* is competing with bee *1*, potentially driving *1*'s gain down from 1.0 to 0.5 units, while increasing its *own* gain only from 0.4 to 0.5 units. The two bees would then, as a result of their individual optimization, provide the colony with a net gain of only 1.0 units, rather than the 1.4 units obtained through cooperative foraging.

In reality, cooperative foraging would only provide a temporary advantage, and only for the special case where all bees confined their foraging to a small foraging area with few flowers, from which bees from other colonies are excluded. If bee *2* can reduce the foraging profits of bee *1* from 1.0 to 0.5 units, then this means that flower numbers are low. *Both* bees will have to go to flower A anyway, to pick up the remnants, sharing the 0.5 units between them.

Bumblebees do not monopolize feeding territories, and they may forage miles from their nest, as honeybees do. In one experiment it was shown that twenty-five colonies placed near another colony had little, if any, effect on that colony's foraging success (Sydney Cameron, personal communication). I have seen bumblebees on a mountaintop in Maine with milkweed pollinia on their legs, even though there was thick forest on the mountain and no milkweed could grow there. Bumblebees regularly fly over open ocean in Maine to forage on the offshore islands.

Because of the long distances that the bumblebees will travel to forage from flowers, it is unlikely that one colony by itself can significantly decrease the available resource base in its foraging area. The flowers that are available to it will be emptied primarily by bees of other colonies. Bees foraging only 100 yards (91 meters) from their nest have 6.5 acres available for foraging, and even if the average flower density is no more than one flower per square meter, they still have 26,000 flowers available to them. Those foraging one mile (1,608 meters) from the nest have 2,011 acres available, in which to search for eight million flowers. Those foraging three miles from the nest have 18,103 acres and 73 million flowers available. A bee visiting 20 flowers per minute could not significantly affect future profits of a colony-mate, unless by some chance it happened to visit exactly the same individual flowers. There is nothing to be gained for the colony through cooperative foraging—having some members visit suboptimal flowers—because any rewards that might accumulate in more rewarding flowers are more likely to be taken by bees from other colonies than by bees of the same colony.

Honeybees, which have evolved in the Old World tropics—a resource environment very different from that of the bumblebees—provide an interesting contrast in colony foraging optimization. In the tropics, rich and clumped food resources are commonly available from massively flowering trees, while in the tundra and temperate habitats of the bumblebee, small clumps or individual flowers tend to be scattered diffusely over large areas. Clumped resources, provided they exceed the exploitation capacities of single workers, can be exploited efficiently by those bees that can communicate the food location to hive-mates; the discoverer of a rich resource can confer great foraging success on its hive-mates, without significantly reducing its own foraging success.

Unlike bumblebees, honeybees rely heavily on communication as part of their overall foraging strategy. Having discovered a profitable resource (which may not be the best available) the scouts begin to advertise their find in the hive by "dancing." The dance is a symbolic la guage that tells recruit bees to leave the hive and search for food with a scent like that deposited on the dancer. When the food source is distant the dance also provides information on direction from the hive —relative to the sun's position—as well as distance. Potential recruits follow the dancers, "read" the message, and then search where they have been directed. (Karl von Frisch spent a lifetime unravelling the details on the dance language of bees, which he describes in his books *The Dancing Bees* and *The Dance Language and Orientation of Bees*.)

Many honeybees may be simultaneously discovering a wide variety of food sources in many areas. Confusion of the hive by excess data is prevented by a series of behavioral mechanisms that act to produce a consensus on the best available resources. First, some confusion is avoided in that the number of scouts is small and only a small percentage of the foragers dance. Second, scouts generally do not recruit until after making several foraging trips to test the quality and desirability of a resource, unless it is obviously of high quality. Third, the dances themselves, by their vigor, indicate overall foraging profitability, in addition to distance and direction, thus providing a threshold that may or may not be sufficient to alert new bee recruits. Fourth, the recruitees are given food samples by the dancer, from which they can presumably gauge not only the kind of food source, by its scent, but also its quality, by its sugar concentration. Fifth, foragers get rid of their nectar (but not pollen) by regurgitating it to receiving bees in the hive, and they do not dance if they cannot get rid of their foraging yield. Thus *tanzlust,* or dance enthusiasm, depends on hive demand for the product advertised. The receiving bees indicate hive demand by taking nectar from those returning foragers that have the most concentrated nectar rather than from those that bring dilute nectar (unless the colony is in need of water to prevent overheating at high air temperatures). When returning foragers find no takers for their collected nectar, they stop foraging and dancing. The various decisions in the communication process involve the whole hive, so that hive response is channeled to the best resources currently available.

Various attempts have been made to put the honeybees' ability and behavior to practical use by training them to visit selected crops (Free, 1958). The usual method has been to have bees returning to the hive walk over a bed of picked flowers of the kind that one wishes them to visit. The successful foragers returning from *all* crops would thus have the same scent adhering to them—scent from flowers they may not have visited. In a sense, the bees are made to lie about the flowers they visited to collect their foraging loads. The results have been mixed, probably because bees cannot be fooled for long if they do not get adequate food rewards from the flowers indicated.

While a honeybee colony relies on cooperation and communication for success in foraging, a bumblebee colony relies on individual initiative. Each is suited to a specific life history strategy in a particular environment. A honeybee colony resembles a big corporation that goes after the big markets. It can wait out long lean periods and rapidly exploit windfalls because of its elaborate organization and huge communal storage capacities. In contrast, a bumblebee colony has a more individualistic cottage-industry approach. It thrives by living hand to mouth, exploiting small, scattered energy sources that in most cases can be taken up by single workers operating individually. Windfalls, such as large banana inflorescences loaded with nectar, are rare for bumblebees but not for honeybees, which originated in tropical areas. Bumblebees do not need an elaborate communication system. In fact, it could be disadvantageous, for it could reroute bees who might be better employed in seeking their own food sources. Rather than loafing in the hive, and waiting to follow dances, the bumblebees keep track of changing resources through their minor specialties, and they explore for new food resources when their majors no longer satisfy.

The strategies employed by honeybees and bumblebees, respectively, offer interesting analogies to human systems. Humans read their customs and ideas into nature, and they attempt to find confirmation for them in the "natural" order. But in doing this we should keep in mind that "No good sensible working bee listens to the advice of a bedbug on the subject of business," as Elbert Hubbard wrote. Insects can tell us nothing about human problems, and we should not look to the bee, nor the bedbug, for examples. It would be as mad to try to fit us, as rational beings, into situations where we would have to act like automatons and abrogate reasoned choices, as it would be to make the bees' actions contingent on reason. Yet bees have been taken as exemplars of all kinds of "natural" orders since biblical times. Bee hives have been seen as monarchies, with a queen and her loyal obedient subjects, or as communistic organizations, where each labors according to his or her ability, and each receives according to need. In the economics of the bumblebee colony, and particularly their organization of foraging, one can perceive the operation of individual motivation, as each individual bee tries to optimize its foraging success. And this success results in the good of the whole colony, as if the individuals were led by an "invisible hand." In this sense a bumblebee hive bears some interesting resemblances to the economic model outlined by Adam Smith in *Wealth of Nations*. Smith proposed that "the uniform, constant, and uninterrupted effort of every man to better his condition, the condition from which public and national, as well as private opulence is originally derived, is generally powerful enough to maintain natural progress of things toward improvement." Smith observed further that "It is not out of benevolence of the butcher, the brewer, or the baker that we expect their services but out of the care for their own interests." He felt that individual initiative was the most potent force for public good. In bumblebee society, other things being equal, those colonies whose foragers exercise the most individual initiative in finding and skillfully exploiting the most rewarding flowers will be the ones producing the most new queens and drones.

The allocation of foraging specialities in bumblebees, resulting in the specialization of individuals and consequent advantage to the colony, is also analogous to Smith's concept of specialization in human societies. Smith argued that individuals would specialize only where their labors would generate profits, and these profits would be exchanged (by money or other capital) to the benefit of others in society with other skills. Specialization, in turn, greatly improves productivity, since no one individual then need master all the skills to provide all of life's needs. Everyone benefits as the goods and services are exchanged throughout society. Individuals, with their own advantage in view, are employed in positions (or at flowers, in the case of bees) most advantageous to society. Eventually, a division of labor results that fits society's requirements (or wants). Among bumblebees, unspecialized individuals do not reap potential rewards as rapidly as specialists, and no bees specialize on flowers with low rewards if they can find better ones.

Smith also proposed that, in a society, specialization necessarily results in interdependence. And the interdependence (and specialization) can be achieved only through an exchange of accumulated capital (representing labor). This is also true for bumblebees, except that the resources collected are immediately accessible to the whole community, rather than going first to individual pots before feeding back to society. For example, those individuals that collect only pollen feed on honey collected by others, and foraging specialists leave the hive duties and reproduction to others. In social bees, of course, the capital exchanged among different specialists in the colony is honey and pollen.

Capital represents labor and derives its value from it. For bees the precise value of their capital, honey, can be easily calculated from the labor they invest to collect it. For example, I observed some *Bombus fervidus* visiting forty-four red clover blossoms per minute; since the blossoms contained, on the average, 0.05 mg sugar (when they had been screened to exclude competitors) the bees were taking up to 2.2 mg sugar per minute. Their foraging investment was, on the average, 0.1 mg sugar per minute (see Chapter 7), so that a typical bee was accumulating 2.1 mg sugar during each minute of foraging. Bumblebee honey is about 90 percent sugar, as is honeybee honey. Thus, one pound of honey (0.45 kg) contains 408,000 mg sugar. Thus, one pound of honey from red clover represents, for *B. fervidus*, about one bee-year of foraging labor, assuming ten-hour work days ($1/408 \times 10^3$ mg $\times$ 2.1 mg/min $\times$ 600 min/day $\times$ 365 days/yr = 1.13 bee years); during that time the bees will have visited and possibly pollinated 9.6 million clover blossoms. On the open marketplace of the fields, however, where bees are not artificially excluded from flowers, blossoms contained on average only 0.005 mg sugar, and the value of the same amount of capital is actually 11.3 years of bee labor, rather than 1.13.

Bumblebees with short tongues, and honeybees, visit less than half as many clover flowers per minute as *B. fervidus,* and the cost of their labor is so prohibitively high on the open marketplace that they are outcompeted and replaced by the more rapid workers like *B. fervidus.* The short-tongued bees, in turn, can outstrip *B. fervidus* on other flowers, particularly those with short corolla tubes.

As in the economic system described by Smith, there is seldom inflation in the bee's system—a given amount of honey (capital) always represents a more or less constant amount of labor. This is because strikes, used in the human system to increase by force the value of labor, are impossible. In the bee's world scabs are legion, and the value of labor cannot be increased artificially (unrelated to effort required to make the product) because there is no organized force to eliminate competition among labor.

Neither is there tyranny from above. There are no monopolies controlling any given resources. All of the bees have access to all of the flowers.

CHAPTER TEN

Where the bee sucks,
there suck I.
—Shakespeare, *The Tempest*

Competition between Species

A glance around a bog or meadow may reveal dozens of bees moving
in apparently random fashion from flower to flower. A closer look
would show that the bees are of different species, and each species
includes bees from different hives. Given this information, a close
study of the movements will show less randomness. All of the bees are
looking for bargains in nature's supermarket. Some are finding them.
Different species may favor different flowers. Different individuals
from the same hive may occupy specific foraging areas and visit specific individual plants. Some bees leave a flower when approached by
another bee, or will not visit a flower that has been recently utilized.
Our first casual glance reveals the obvious: many bees are utilizing the
same resources. What is less obvious is that there is intense competition for the few transient bargains, and a variety of strategies can be
applied in competing for them.

The flowers in an area are never available to only one bee. There are
usually many bees utilizing the same resources, and to determine the
best method for acquiring the resources a number of decisions must be
made. One of them is, should all the effort be spent in foraging itself, or
should time and energy be spent patrolling the area and attempting to
repel competitors from it? The latter, called contest or interference
competition, is observed in many kinds of animals holding territories
containing rich and tightly clumped resources. A good example of this
strategy is found in a few species of stingless bees in Central America.

Heaps of dead bodies are sometimes left as a consequence of aggressive encounters between competing colonies of stingless bees at superior food sources, such as banana inflorescences (Johnson and Hubbel, 1974). Those colonies that have the most aggressive bees, and which are numerically superior, dislodge others by force from the best available food resources.

Stingless bees, like honeybees and bumblebees, belong to the family Apidae. They have been used for centuries by the indigenous peoples of tropical America as a source of honey and beeswax. Most species of stingless bees are able to recruit nestmates efficiently and arrive in crowds of hundreds at a rich food source, such as a sugar bait or the inflorescence of a tropical flower, that has lots of pollen and nectar. This efficient recruitment benefits the stingless bees in at least two ways. First of all, high quality food resources can be rapidly exploited by a colony; they might otherwise be harvested by competitors. The crowds of bees can defend a good food source against new invaders, particularly scouts. It is important to prevent scouts from discovering the superior food, for if they do discover it they may return with large numbers of recruits and by their combined efforts dislodge their competitors.

The aggressive strategy of some species of stingless bees is advantageous at rich and highly clumped food, but it can generally be employed only by large bees that can efficiently recruit large numbers of nestmates to inflict costly damage on their competitors. The bees of some of these species invariably attack other bees holding superior food—but only if they have a chance of eventually securing the resource without much damage to themselves. They act reliably according to their own best interests.

There is no set formula for the best foraging behavior. For example, the optimum response changes drastically when food resources become less compact. The wider resources are scattered, the less efficient it is to recruit and defend specific items, and the more difficult it is to patrol and defend an area. Competitors then appear to work peacefully (without contact) side by side, but they may still compete relentlessly by trying to remove resources faster than the next individual. Aggressive encounters then become a liability, for even the winners lose—they have only expended time and energy that could have been used for foraging. The nonaggressors, which do not interrupt their foraging, reap more food energy and are competitively superior. Such competition, called scramble or exploitation competition, generally results in the depletion of resources to the very minimum of economic profitability. In turn, it selects for energy economy and foraging efficiency in the contestants.

Bumblebees generally do not have rich, clumped resources available to them. The nectar contents of most plants in temperate and arctic regions are minute, and flowers are scattered widely. Bumblebees must visit hundreds of flowers on each foraging trip. It would probably not be advantageous for individuals to defend feeding territories, since each foraging trip involves searching over a wide area, sometimes 500 square meters or more. Furthermore, bumblebees, unlike honeybees and stingless bees, are unable to recruit nestmates to specific sites containing food. Apparently, given their resource environment, there is no advantage for them in seizing, holding, and defending territories, either as individuals or as colonies, and the communication and fighting capacities to make such a strategy possible have not evolved.

Bumblebees of several species routinely forage side by side, sometimes within millimeters of each other. Unless they are within extremely short range of one another they usually give no sign of intolerance while foraging under natural conditions. Many hundreds of individuals from many species and many colonies commonly share the same flowers in the same foraging areas. Competition between bumblebees is therefore primarily exploitative, in that each individual, while foraging at the flowers where it can make the most immediate profits, leaves less food for others at that particular source.

Recently, Douglass H. Morse (1977), an ecologist who has switched his studies from wood warblers foraging in high trees and dense thickets to the more amenable bumblebees in Maine meadows, has reported on possible interference competition among bumblebees foraging on the same flowers. He concludes that small bumblebees, such as *B. ternarius*, forage at the tips of goldenrod inflorescences and hypothesizes that this is because they want to avoid the large bees, *B. terricola*, that are utilizing the center sections. However, it is not yet clear that foraging worker bees distinguish each other according to species, nor that they accurately measure their own size in relation to that of bees foraging near them. They have no need of such behavior. They never engage in physical contests, in the field, and accurate assessment of the size of competitors is of dubious advantage. Possibly smaller bees (*B. ternarius*) can forage further out on the slender drooping branchlets of the goldenrod inflorescences than heavier bees (*B.*

terricola) can. The tendency of *B. ternarius* to feed more distally in the presence of *B. terricola* could be, at least in part, the result of a preference for feeding in undepleted areas (those that *B. terricola* could not exploit), rather than, or in addition to, active avoidance of larger bees.

In late summer, when bumblebees are numerous, food rewards are allocated on a "first come, first served" basis. Bumblebees sometimes work from dawn till shortly after dark, visiting isolated flowers at a sustained rate of 20–40 per minute, generally without pause. I followed one worker for an exhausting (mainly on the eyes) two hours and two minutes on a single foraging trip during which it visited 454 goldenrod panicles (each panicle consists of many hundreds of tiny inflorescences and each inflorescence is composed of dozens of near microscopic florets) and 329 aster inflorescences, as well as 19 jewelweed blossoms. I was unable to keep track of the probably many thousands of tiny individual goldenrod inflorescences and aster florets the bee visited during that time. Another bee visited at least 800 jewelweed blossoms on one foraging trip of about two hours. Obviously, it was not easy for the bees to make quick energy profits, because the nectar was being depleted rapidly by competitors. Nectar abundance was observed only in early summer, when most foraging trips were less than 20 minutes in duration.

Recently Chris Plowright and Bruce A. Pendrel have examined competition for pollen resources among bumblebees in New Brunswick, Canada. From records of the amounts of pollen brought in by individual workers in *B. terricola* colonies, they showed that there was a decline in pollen foraging success throughout the season, as bumblebee populations were building up. Other evidence for competition (although from only a small sample of colonies) comes from a vast unplanned ecological experiment, which started in 1949 when DDT was sprayed over millions of acres of forests in Maine and New Brunswick to kill spruce budworm, a caterpillar that was undergoing a population explosion. Spruce budworm feeds on and kills the spruce and balsam fir used by the pulp and paper industry. Normally, the spruce budworm moth population explodes in periodic outbreaks; each lasts several years, but the outbreaks occur only every thirty-six years. The body count showed that DDT killed millions of caterpillars. But at the same time DDT application maintained the plague indefinitely by keeping the population cycle from going to completion, since diseases, parasites and predators were not allowed to build up. The budworms were still abundant in 1968, when DDT was discontinued (because it was noticed that it caused large bird and salmon kills). In 1970, a biodegradeable organophosphate nerve gas, Fenitrothion, was substituted for DDT. This substance was also sprayed over millions of acres. Fenitrothion is highly toxic to bees, unlike DDT, which is only moderately toxic to them. Almost immediately after the new poison was applied to the forests, growers saw no more bees on their blueberry heaths, and crops failed near the spray areas. Undoubtedly the forest insects were equally affected, upsetting seed and food production for animals higher up in the food chain (Kevan, 1975).

Meanwhile, on the blueberry fields and in millions of acres of forest, the competition for pollen and nectar among the few surviving pollinators was reduced to a minimum, providing ideal conditions for an ecological experiment that could not normally be perpetrated. Plowright and Pendrel introduced *B. terricola* colonies into areas that had been sprayed, and they compared the foraging success of these colonies in sprayed areas with those in unsprayed areas that contain the normal population densities of bees. They found that the pollen input to a colony per minute of foraging time was five times greater in the sprayed than in the unsprayed area. But the results of these and similar experiments would be expected to vary greatly from one part of the season to another, and from one locality to the next, as well as from year to year; food is never uniformly scarce.

Low rates of food intake to a colony, due to competition among foragers when and where food is scarce, undoubtedly contributes to colony growth rates that are lower than the physiological maximum and to early colony failure, as well as to delay, reduction, or absence of production of new drones and queens at the end of the colony cycle. However, the effect is not necessarily a direct one. To some extent the bees compensate. Colonies often maintain some foraging capacity in reserve. When food supplies are low, some bees that might otherwise have become house bees become foragers, and individual foragers probably increase their foraging. However, greater foraging effort by the colony is not without cost—it results in greater worker mortality, and it leaves fewer bees to guard and regulate the temperature of the nest.

I have so far treated bumblebees as if they were all alike. In reality there are individual and species differences. And this makes our model yet more complex. Workers of a colony differ in size, and larger bees have longer tongues. Partly because of these morphological differences, different individuals are better able to forage at different flowers. The tongue is the primary foraging tool.

The tongue is long, slender, and hairy at the tip (Fig. 10.1). It terminates in a small lobe that is probably used as a suction cup—the bee creates the suction with a powerful pump activated by muscles in the head. The small hairs at the tip of the tongue may also take up, by capillary action, the minute amounts of nectar that the bees collect from the miniscule florets of some flowers. Tongue length determines whether or not, and how fast, a bee can manipulate a particular flower to extract nectar. Long-tongued bees spend less time at each long-corolla flower than short-tongued bees; they reach the nectar without having to wedge themselves far into the flower. Spending less time per flower, they can visit more flowers per unit time (Fig. 10.2). On short-corolla flowers, on the other hand, a long tongue can be a liability, and short-tongued bees can extract the nectar more quickly than long-tongued bees. Long-tongued and short-tongued bees can coexist because they specialize in different flowers.

Bees probably learn by experience to visit those flowers best suited to their tongue-length (Hobbs, 1962). Long-tongued bees generally forage from long-corolla flowers, and short-tongued bees from short corolla flowers (Fig. 10.3). This behavior is significant not only for the division of resources between different species (see Brian, 1957), but also for the division of labor within a colony. The bees in a colony ordinarily have tongues of various lengths and as a result can tap a wide range of food resources.

The abundance and low species number of bumblebees probably results from their ability to utilize many kinds of flowers. Being able to forage from, say, two equally rewarding types of flowers, a species can potentially build up to twice the number or biomass that would be possible if it were restricted to only one type of flower. The alternative is for the two resources to be utilized by two species, each specializing in one of the resources. The greatest number of species, and the lowest biomass per species, would occur if each flower species were served by one specialized foraging species. To some extent, solitary bees specialize in specific flowers, and this group is extremely rich in species numbers. In the eastern United States alone there are nearly eight hundred species of solitary bees. In the southwestern United States, with its ephemerally blooming desert plants, the number of solitary bee spe-

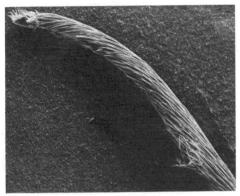

Fig. 10.1 Scanning electron-micrograph of the terminal 3 mm (right) and the 0.2 mm tip (left) of the tongue of a bumblebee worker. Courtesy of David W. Stanley.

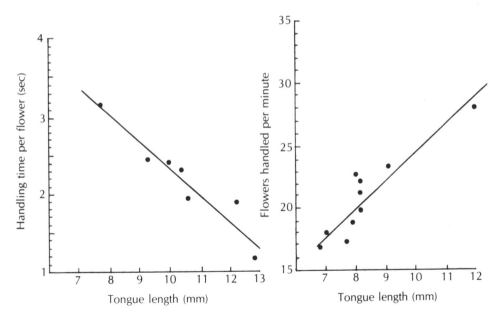

Fig. 10.2 Relationship between tongue-lengths of bumblebees from different species and their working speeds at two different kinds of flowers with deep corollas. Data on handling time per flower refer to *Delphinium barbeyi* (adapted from Inouye, 1977). Data on flowers per minute refer to red clover (adapted from Holm, 1966).

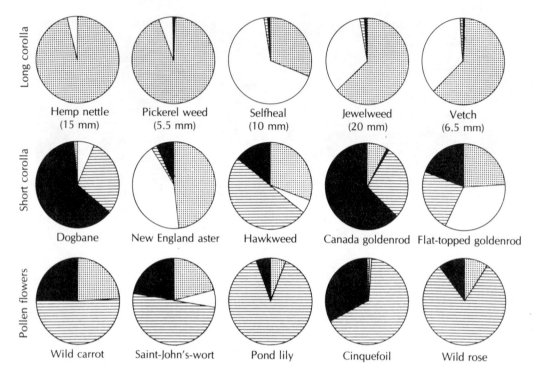

Long corolla

Hemp nettle (15 mm)   Pickerel weed (5.5 mm)   Selfheal (10 mm)   Jewelweed (20 mm)   Vetch (6.5 mm)

Short corolla

Dogbane   New England aster   Hawkweed   Canada goldenrod   Flat-topped goldenrod

Pollen flowers

Wild carrot   Saint-John's-wort   Pond lily   Cinquefoil   Wild rose

Fig. 10.3 Partitioning of long-corolla flowers (top row, with corolla depth in parentheses), short-corolla flowers (second row, corolla less than 3 mm deep), and pollen flowers (bottom row) among four common bumblebee species (stippled = *B. vagans,* open = *B. fervidus,* lined = *B. terricola,* solid black = *B. ternarius*) in one area near Farmington, Maine, in 1973. All of these flowers were visited primarily by bumblebees. Although general trends in partitioning are apparent, the exact utilization of the different kinds of flowers varies from year to year and locality to locality, according to the local relative abundance of each short- and long-tongued species. (Adapted from Heinrich, 1976c.)

cies is much greater, and many of these bees are morphologically and behaviorally adapted to utilize the pollens only of certain specific plants (Linsley, 1958). The many solitary bee species can coexist because they divide up the resources in this way. But the bees are restricted to those areas and to those particular times that their host plants are in bloom.

Most bumblebee species, in contrast, have relatively wide distributions. Many are found across the entire United States. These bees are not restricted to the utilization of particular plants, and hence interspecific competition is probably intense among bumblebees, as well as between bumblebees and other bees. There are only sixteen social bumblebee species in the eastern United States, and fifty in the entire country.

When Charles Darwin studied the evolution of the Galapagos *Geospiza* finches he was struck by the differences in bill morphology, and he attributed the evolution of these differences to adaptive radiation. The various bill adaptations allowed the utilization of various food resources. The birds had specialized their mouthparts through evolution and thus had partitioned their resources. The same phenomenon is well-known among bumblebees. As already mentioned, tongue-lengths vary with body size within a bumblebee species and even within a colony. These differences partly determine which flowers individuals specialize in. The tongue-length differences between individuals, however, are not as large as the tongue-length differences between species, which affect the ranges of flowers utilized by particular species. The differences between species have probably evolved to reduce interspecific competition. In general, long-tongued bumblebee species (having tongues that are 75–85 percent of body length) utilize long-corolla flowers, while short-tongued bumblebee species (having tongues that are 50–60 percent of body length) tend to be restricted to short-corolla flowers. Two or more species of bumblebees with tongues of similar length may often forage from the same flowers. If they rely heavily on these flowers for food and thus greatly depress the food rewards available, then it is likely that, as a result of competition, the more efficient foragers will replace the less efficient. I found a possible example of competition in action in a jewelweed patch in my study area in Maine. In 1973 the flowers were utilized almost exclusively by two long-tongued bumblebees, *B. fervidus* (Fig. 9.3) and *B. vagans*. These bees were very numerous and removed all the nectar the flowers provided. The shorter-tongued bees, *B. vagans* (8 mm), were most active early in the morning while it was still cool (Fig. 10.4). The longer-tongued bees, *B. fervidus*, started to forage later when it was warmer and the cost of thermoregulation was less. Possibly *B. fervidus* did not need to forage early, because they could still find nectar after *B. vagans*, with its 3 mm shorter tongue, could no longer reach any. In midday *B. fervidus* were very numerous, while *B. vagans* stopped foraging at jewelweed, presumably because they were no

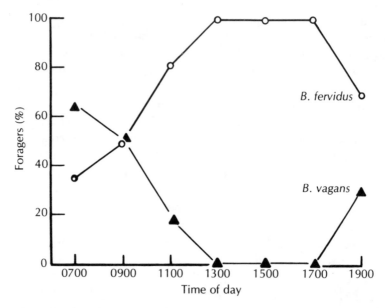

Fig. 10.4 The relative abundance of two kinds of long-tongued bumblebees at a jewelweed colony from morning till evening on the same day in late August 1973. There were few or no other foragers. The same colony was also observed in August 1974 (see Fig. 10.5). (From Heinrich, 1976c.)

longer finding rewards. In other years when *B. fervidus* were not numerous the *B. vagans* foraged at jewelweed of this patch throughout the whole day. (*B. fervidus* were rare in the years after 1973, but the reasons for this decline in their abundance is not known.) The absence of *B. fervidus* at the jewelweed flowers in 1974 possibly affected other bumblebee species (Fig. 10.5). The nectar left by *B. vagans* in the recurved nectar spurs was harvested by the short-tongued bees, *B. terricola* and *B. ternarius*. These bees were robbing the flowers by biting into the nectar spurs and taking the nectar without entering the flowers (Fig. 9.4).

Only three or four species of bumblebees are likely to be abundant in a small locale, such as a meadow, bog, or mountaintop, although a dozen or more may be present in the larger surrounding area. Generally, one of the abundant species is short-tongued, another is long-tongued, and a third has a tongue of intermediate length (Fig. 10.6).

There are no *Bombus* in Europe with tongue lengths as short as those found in North America. But honeybees, which are native to Europe and not North America, have tongues that correspond closely to the shorter tongues of North American bumblebees (Inouye, 1977).

Coexisting species also partition another important, and sometimes limiting, resource: nest sites. Some species nest in the ground, others *on* the ground, and still others nest in trees. For example, in Maine I have found nests of *B. perplexus* in bird houses and in the walls of our barn and house. *Bombus fervidus* nests are usually located on the surface of the ground, but sometimes they are underground. *Bombus terri-*

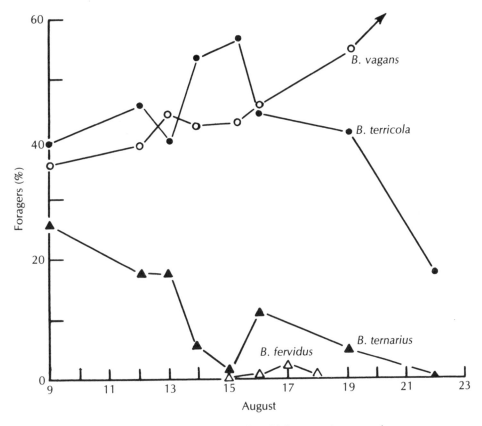

Fig. 10.5 The relative abundance of different bumblebee species at a colony of jewelweed in Maine during August 1974. Foragers other than bumblebees were rare. (From Heinrich, 1976c.)

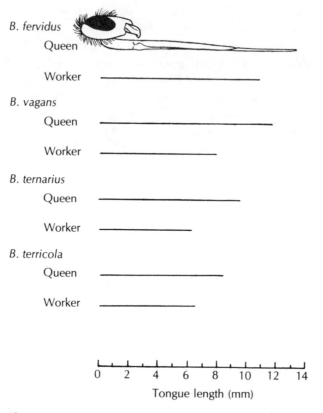

Fig. 10.6 Average tongue-lengths of the queens and workers of four common bumblebee species in Maine. (Adapted from Heinrich, 1976c.)

cola nests are usually underground, but I found one colony in a bird house.

The intensity of competition among bumblebees is indicated by the full use to which resources are put each year, regardless of how sparse the bumblebee population may be in early spring. During some years the overwintered queens are extremely abundant in the spring. Hillsides overgrown with blueberries, and leatherleaf patches in bogs, are audibly conspicuous because of the continuous humming of the many queen bees foraging in May. In other springs, however, there are very few queens to be found in the field, possibly because of a high mortality, resulting from deep frosts in the winter of flooding after the thaw. Nevertheless, regardless of the number of queens about in spring, in late summer and fall there are usually very large number of workers and drones foraging on asters and goldenrods. The bees generally utilize most of the available nectar in late summer, regardless of initial queen population density. It is probable, therefore, that when overwintering success is poor, the surviving queens tend to be successful in establishing large colonies. But when overwintering success is high, the new queens—hundreds may survive from a single colony—quickly utilize the available food resources, which limit the number and size of colonies.

The competition for food is, of course, most intense between colonies of the same species and between species with similar requirements. However, bumblebees also compete with other wild bees, as well as with the familiar honeybee, a European import. The small wild bees generally reduce their competitive overlap with bumblebees on the same flowers; they do this by foraging at higher temperatures and by utilizing flowers that have been already visited by bumblebees. Their relatively low energy demands may allow them to get by with leftovers below the feasible economic foraging threshold of bumblebees. I have frequently observed large numbers of solitary bees foraging near midday, long after the bumblebees had depleted some flowers to the point where they no longer contained visible amounts of nectar.

The social honeybees, like the bumblebees, range widely in their foraging. Bumblebees visit all the flowers visited by honeybees, and honeybees are excluded only from rare flowers (such as turtlehead and closed gentian) visited by bumblebees. Generally, both types of bees forage at the same time. Because of the almost total overlap in resource utilization between honeybees and bumblebees, the competition between them is probably intense. In Maine I have seen relatively few bumblebees in those areas where there were many honeybees. Up until 1975 honeybees were rare or absent from most of my study areas. In the fall of 1977, on the other hand, the goldenrod was overrun with honeybees. There were several honeybees on almost every panicle, and only rarely a bumblebee. In previous years it was the other way around, with bumblebees being far more numerous. I inquired around the neighborhood and discovered that beekeeping had become a new hobby. In the previous two years, 15–20 new hives had been established by various neighbors within two miles. Each hive yielded a surplus of about 50 pounds (22.7 kg) of honey.

Honeybees were originally imported by the early settlers from Europe. (Indians called honeybees the "white man's flies.") Men have provided food for them by clearing the land, cultivating crops, and introducing many weeds. However, when honeybees are present in areas with native vegetation, where wild bees normally harvest nearly all of the available nectar and pollen, they adversely affect the populations of bumblebees and other wild pollinators. Decreases in wild bee populations appear to be directly proportional to the amount of nectar and pollen made unavailable to them, although at present there are no data to confirm or refute this hypothesis.

The potential effect of the competitive pressure can, however, be deduced from a rough calculation. For example, as I mentioned previously, a colony of *Bombus vosnesenskii* containing 239 larvae and 136 adult queens utilized daily about 54 g honey and 25 g pollen. If this was an average daily food consumption for a colony rearing 375 reproductives (239 + 136), then each was directly or indirectly requiring .14 g honey (54/375) and 0.7 g pollen (25/375) per day. Egg-to-adult development takes about a month, and the young queens stay with the nest for at least two more weeks; thus during her development each new queen consumes about 6.3 g honey (45 × .14) and 3.15 g pollen (45 × .07). The rearing of 375 reproductives may thus require about 2,360 g of honey and 1,181 g of pollen. This is a minimum estimate, because it does not take into account the food resources required to build the "machinery" (the workers) that are used to make those reproductives. Nevertheless, even if the bumblebee colony loses to honeybees only that food required near the end of the colony cycle (when there is the most competition for food), then for every gram of honey produced by honeybees in an area suitable for bumblebees there would be about .16 (1/6.3) fewer bumblebee reproductives produced. A strong hive of honeybees in the United States collects on the average about 200 kg of honey for its own use, and about 40 kg surplus that can be collected by the beekeeper. A single honeybee hive, then can reduce the population of bumblebee reproductives by 38,400 (240,000 g × .16).

CHAPTER ELEVEN

To make a prairie it takes a clover and one bee,
One clover, and a bee,
And reverie.
—Emily Dickinson, *Prairie*

Pollination and Energetics

The bumblebee's ecological puissance was noted by Charles Darwin and his contemporaries. Darwin discovered that red clover required bumblebees for its pollination. This led the German biologists Karl Vogt and Ernst Haeckel to propose, facetiously, that the British Empire owed its power and wealth to bumblebees, since its power resided mainly in its Navy, which subsisted on beef, which came from cattle that had fed on clover, which was pollinated by bumblebees.

The story is actually more complex. The celebrated British biologist Thomas H. Huxley allowed that the success of the British Empire ultimately depended on spinsters: spinsters kept cats, which kept down the mouse population; mice destroyed bumblebee nests, so as mice became fewer, bumblebees increased, which ultimately profited the Navy. In fairness to married folk, one could present a competing hypothesis: bumblebees need mouse nests to house their colonies, so without spinsters and without cats the bees would have ample nest sites, and their populations could increase to pollinate clover and bring beef to the Navy, etc.

The exact role of spinsters in the British Empire can be debated, but bumblebees pollinate thousands of plant species besides red clover and they have an ecological impact beyond their role as perpetuators of clover.

Bumblebees visit very nearly all rewarding and some nonrewarding flowers in their habitats. They also pollinate most of these flowers, but only a small spectrum of the thousands of flowers they visit are adapted to them exclusively. The majority of plant species are pollinated by a large number of insect species. The relative abundance of each pollinator varies not only diurnally, seasonally, and from one year to the next, it also varies from one locality to another. For example, in Maine on June 11–13, 1972, bumblebees were nearly the exclusive visitors of blueberry blossoms on Mt. Tumbledown, although at three other habitats blueberries were visited by large numbers of other foragers, primarily solitary bees (Fig. 11.1). The species composition of the bumblebees also varied from place to place. In forest, near Kokadjo, by far the most numerous visitors to blueberries were *B. vagans*. In a nearby field, the most common visitors to blueberries were *B. terricola*. In Huckleberry Bog there were relatively few *B. terricola* but many *B. ternarius* and *B. perplexus*. On Mt. Tumbledown both *B. vagans* and *B. terricola were common. The blueberry blossoms are adapted* for pollination by both short- and long-tongued bumblebees, as well as solitary bees and possibly other insects. By being simultaneously adapted to utilize many different kinds of pollinators a plant is hedging its bets, and can produce seed despite the unpredictable changes in pollinator availability in different parts of its range.

Because most flowers have coevolved with a spectrum of pollinators, it is nearly impossible to separate the very important ecological role of bumblebees from that of other bees, and some other insect pollinators. What bumblebees do, other bees do, and vice versa. If one restricts attention to those few instances where only bumblebees act as pollinators, one severely slights their actual role in pollination.

Bees are of inestimable importance in crop pollination (Free, 1970). It has been estimated that in North America, humans are ultimately dependent on bee pollination for up to one-third of their food. Bumblebees are worth billions of dollars to North American farmers for clover pollination alone. Bees pollinate more than 2,500 species of crop plants and they ensure seed crops of innumerable species of native plants that maintain ground cover, prevent erosion, and feed animals. Their activity is thus vital, not only for our food supply, but also for land, water, and animal resources.

Because of its ease of management, the common honeybee, *Apis mellifera,* is our most important pollinator of commercial crops. But the value of honeybees is severely limited at air temperatures less than 15°C. They are also of limited value on crops having flowers with deep, narrow corollas (such as broad beans and red clover), or having

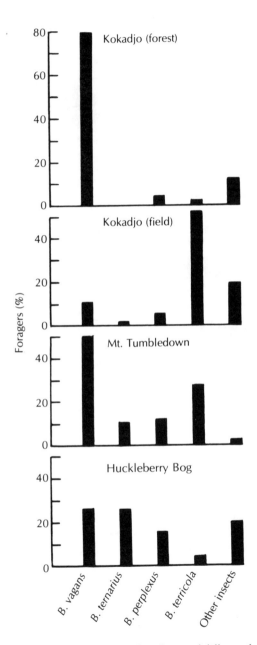

Fig. 11.1 The relative abundance of different foragers foraging from blueberry blossoms at four different sites in Maine during June 11–13, 1972.

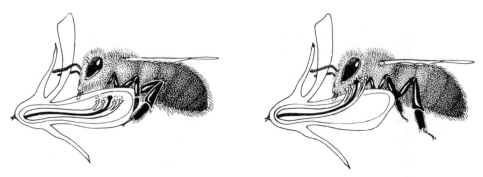

Fig. 11.2 Pollinating mechanism in alfalfa flowers showing untripped (left) and tripped (right) flower. In this spring-loaded system, the pistil and anther filaments are held under tension in the lower lip of the flower.

special mechanisms of pollen release, such as alfalfa. In alfalfa flowers the male and female organs are held under springlike tension by the lower part of the flower, known as the keel. When a bee enters the alfalfa flower, the bottom of the flower, the keel, sags under the bee's weight, activating the trip mechanism that releases the male and female organs (Fig. 11.2). The organs hit the bees on the lower portions of the head; pollen from the flowers' male organs are deposited on the bee, while the female organ of the flower becomes dusted with pollen the bee has picked up at a previous flower. Honeybees apparently do not like having their heads snapped at each flower visit. They learn to enter at the side of the flower to avoid tripping it. Untripped flowers are unpollinated and set no seed. More robust bumblebees and some other large bees, however, appear to be less bothered by the trip mechanism, and when they visit the flowers they trip the keel and cause pollination. In addition, they are not limited in their activity at low air temperatures, like honeybees. Furthermore, their long tongues allow them to visit and pollinate clover, which has its nectar sequestered in long, tubular corolla tubes (Holm, 1966). Bumblebees were specifically imported to New Zealand from England between 1885 and 1906 for clover pollination; they also play a dominant role there in the pollination of commercially grown currants, gooseberries, raspberries, broad beans, kiwi fruits, and alfalfa.

It has so far not been possible to maintain large, reliable bumblebee populations in areas of intensive agriculture. Bumblebees need unplowed land with thick matted grass where mice live and leave their nests. They also need an unbroken progression of plants in bloom throughout their long nesting cycle. But solitary wild bees have been used extensively in agriculture, particularly the alkali bee, *Nomia melanderi*, and the alfalfa bee, *Megachile pacifica* (Bohart, 1972). These bees are used commercially in seed production for alfalfa, an important food for the beef and dairy industries. Each individual bee pollinates enough flowers in her lifetime to produce more than 500 grams of seed. Both of these species of leaf-cutter bees are tractable mainly because it is possible to manage their nest sites. Alfalfa leafcutter bees, a hole-nesting species, are kept in alfalfa fields by being provided with shelters in which are placed drilled boards or densely packed drinking straws that serve as nest holes. Alkali bees are native to the western deserts in the United States, and they nest in burrows dug in the moist soil of salt flats. Alfalfa growers protect natural nest sites with fences and build artificial nest sites by recreating the appropriate soil conditions. Honeybees, although they need flowers for several months of the colony cycle, can be managed in agro-ecosystems because when a crop stops blooming one can move the hives to a different area with another crop. In the East, for example, they may be moved from apple, to blueberry, to legume crops.

Few bees and flowers form exclusive one-to-one relationships, but some solitary bees, like the squash and gourd bees (*Peponapsis* and *Xenoglossa*) come close (Hurd, Linsley, and Whitaker, 1971). These bees derive all of their energy supplies from the nectar of *Cucurbitas* (gourds, squashes, pumpkins) and also obtain all the pollen they feed their larvae from these plants. In addition, they mate in the flowers and sleep in them. Generally, the flowers open early in the morning and then wilt. The bees emerge early in the mornings, sometimes before dawn, chewing their way out through the walls of the wilted flowers in time to forage from newly opened flowers that offer nectar and pollen.

In most plants, outcrossing results only if the food rewards of the flowers are in balance with the energy needs of the pollinator. The flowers of a plant must provide sufficient food reward to be competitive with other, concurrently blooming flowers that the bees could potentially visit. But rewards must be low enough so that the pollinator does not confine its visits to the flowers of a single plant (Heinrich and Raven, 1972). Mismatches in energy expenditures of pollinators and food rewards of flowers result in pollination inefficiency. For example, when honeybees visit the giant saguaro cactus growing in our southwestern deserts (McGregor et al, 1959), they tend to limit their visits on successive foraging trips to specific flowers. The flowers, which are normally pollinated at night by large, energy-demanding bats, produce about 5 ml of nectar each (most bee-flowers contain about 50,000 times less). The bees can cross-pollinate the cacti only if their movements transport pollen from anther to stigma of different plants. The bees near Tucson, however, visit only specific flowers on given arms of given cacti. Having all of their energy needs met on the same plant precludes their being of much direct help in cross-pollination.

The site-specificity of bees presents a general problem for pollination of trees with massive blooms (Free, 1960). Recent studies in Costa Rica indicated that most lowland forest trees were outcrossed; flowers fertilized with their own pollen did not set seed (Bawa, 1974). Bees were the major group of pollinators for the trees. In an effort to study their role in pollination, Gordon Frankie and his coworkers (1976) marked hundreds of bees in certain trees with fluorescent dyes and, by capturing bees at the same and neighboring trees, determined intertree movements. They collected and marked some seventy species of bees at one flowering tree, *Andira inermis*. On the first day after marking, most bees remained where collected, but 0.3 to 1.3 percent moved to neighboring trees on the first day. Why did the bees move at all? Many of the trees produced nectar for a very short time—approximately one-half hour in the morning—and this limited time of nectar availability may have caused the bees to disperse. Second, the large crowds of bees at any one tree may have been interfering with each other in contest as well as scramble competition, causing bees to seek other, possibly less crowded, trees.

The making of pollen and nectar to feed pollinators entails an energy cost to plants. But by minimizing the foraging costs to its pollinators, a plant can get away with providing less reward and save energy. There are various ways in which plants can make it energetically less costly for a forager to visit unpollinated flowers. For example, plants with many flowers can maximize a pollinator's visits to unpollinated flowers, and at the same time minimize its energy expenditure for foraging, if they signal to the pollinator which flowers have already been visited (and emptied) and which ones haven't been pollinated and still contain nectar or pollen. Recent evidence indicates that some plants engage in precisely such signaling. Blossoms of some plant species change color markings, scent production, and even geometric outline after being pollinated. There is also evidence that these changes are noted by bees—the bees by-pass flowers that, having been visited, are no longer rewarding and have changed their signals (Jones and Buchman, 1974). Bumblebees restrict their visits, almost exclusively, to newly opened flowers of wild pasture-rose bushes. Day-old blossoms are paler pink, and they were almost completely by-passed.

Plants probably do more signaling than is apparent to human observers. Bees respond to ultraviolet reflection that is invisible to us, and the flowers of some plants change in ultraviolet reflection following pollination. Most of the plant pigments are a class of compounds called flavonoids, and the various types of flavonoids, such as the xanthophylls, anthocyanins, flavanones, and others, are distinguished by simple oxidation changes. These pigments change color through relatively slight biochemical changes, but the biochemical steps between pollination by the insect and pigment change are not known. Furthermore, the phenomenon has been relatively little explored, either from the plant's perspective or from the standpoint of behavior changes of pollinators. I would suspect, for example, that the phenomenon of color change has little functional significance in plants that produce only one or a few flowers, since there would be no disadvantage in a flower's being visited several times. It would be interesting to study the phenomenon of postpollination color or scent change systematically, to see if it occurs more frequently in many-flowered plants, such as trees, than in single- or few-flowered plants. A much simpler strategy, at least in open flowers, is for bees to respond to the nectar content directly. Recently, Robbin Thorp has demonstrated that the nectar in the flowers of almond trees fluoresces under ultraviolet light (Thorp et al. 1975). Almond flowers have a very shallow open corolla with exposed nectar, and by fluorescing when exposed to the ultraviolet of sunlight the nectar advertises itself. Thus, bees can restrict themselves to unvisited, unpollinated flowers on an almond tree.

Scent is another signaling mode. Bumblebees often approach flowers and then veer off at the last moment, as if having decided the flower does not have adequate food rewards. This behavior occurs at white clover, where neither the nectar nor the pollen are visible from the outside of the flower. Which individual flowers do the bees reject, and how do they decide whether or not to land on a flower? On a lawn carpeted with white clover blossoms that were being visited by large numbers of bumblebees, I covered several square meters with bridal veil to exclude the foragers and to allow nectar to accumulate. Outside the screened area the bees rejected a third or more of the flowers, but when the screen was removed none of the unvisited flowers that had been under it were rejected. Could the bees determine by scent whether or not a flower had been visited? I decided to take seriously the aphorism about sniffing the flowers. I lay back in the clover with my eyes closed, and a student held blossoms to my nostrils. With almost no practice I could, with 88 percent accuracy, determine whether or not an individual flower had been previously visited by a bumblebee. Unvisited flowers had a strong, sweet clover scent; visited flowers had a very weak scent.

White clover has minute amounts of nectar. Flowers with large nectar rewards (monkshood, jewelweed, fireweed, or turtlehead) or large pollen rewards (wild rose) were visited in rapid succession by different bees without being rejected by any of them. The high-nectar flowers were, however, unscented even when filled with nectar. Empty flowers of these species were visited just as readily as filled flowers.

A plant must expend energy to produce nectar and pollen, energy that could be budgeted into growth and reproduction. Although the energy from the sun is free to plants, they cannot use it without capturing it and then storing it in the form of sugar molecules. The sugar can be invested to make more photosynthetic tissue, to produce nectar and pollen, or to produce flowers, fruits, and seeds.

If a bee cannot determine how much, if any, food the flower of a particular plant will provide until after it has visited (and pollinated) it, it would seem possible for that plant to cheat. Individual plants growing in crowds could attract bees and be pollinated even if they produced no nectar provided their neighbors did. Indeed, it might be to their advantage not to produce nectar, for then they could invest more energy in seed production. The genes for forgoing nectar production would be advantageous and would spread, for the carriers of these genes would continue to be pollinated and produce more offspring than those plants that produce nectar, because they could divert energy that would have been spent on nectar production into fruit and seed production. However, when the cheaters within the population became too common, the bees would notice that they were making less profit. They would then visit competing plant species that provided greater food rewards (Free, 1968). To some extent, individual plants can get away with not producing rewards, but a species cannot.

The interactions between bees and flowers can be understood in more formal terms by employing the game theory that the British ecologist Maynard Smith has used to describe survival strategies of organisms. Richard Dawkins has very effectively applied game theory to animal behavior in *The Selfish Gene*. The same principles apply to plants. The flower's game is to maximize pollination while minimizing the energy spent on it. The bee's object is to gain maximum rewards from each foraging effort. The bee has to visit large numbers of flowers, thus carrying pollen from plant to plant and ensuring the survival of the plant's genes into future generations. But the genes in one individual plant do not "care" about those in another. We can assume, a priori, that each plant in the population is "selfish" and produces only just enough food to attract foragers and be pollinated. This makes for some interesting juggling of food rewards, for the forager does not treat flowers (or plants) as individuals unless they occur isolated from others. To be attractive the isolated plants must supply large food rewards. However, flowers usually bloom in crowds, and individuals offer different rewards.

How much is enough food reward for a flower blooming in a crowd of flowers? Superficially, from the plant's selfish perspective, it might seem that the answer is "none." This is because a bee may visit, say, a dozen consecutive flowers that have no food reward, and only get paid for its effort on the thirteenth flower, if it has nectar. (Empty flowers may have been visited previously, or they may simply have not produced nectar). In the laboratory foraging arena, with artificial flowers, we have observed bees making up to seventy-five flower visits in a row without receiving a single reward—but only if they had previously been conditioned to flowers of similar appearance.

If bees were to treat plants of a particular species only as a population and never as individuals, cheater genes could spread, leading to the extinction of that species. However, there are opposing selective pressures that save species from extinction by preventing cheating from becoming too widespread. Visible or scented nectar may have evolved as a way of guarding against cheating. In addition, bees that have found nectar rewards become not only flower-constant but also site-specific (Manning, 1956). They concentrate their attention on those groups of flowers that, on the average, have a higher frequency of rewarding flowers than a neighboring group, thus providing a selective pressure for increased food rewards. Evolution measures the relative strengths of the two selective pressures (to provide no food when anonymity is possible and to provide ample food when anonymity is not possible or desirable) and the appropriate rewards are then calibrated with changing conditions over thousands of years.

In order for plants to economize on the expenditure of energy to feed pollinators, animals of low energy-expenditure should ideally be excluded from flowers providing large food rewards (see Fig. 11.3), because the low-energy foragers may restrict their visits to individual flowers, thus tending to eat the rewards without reciprocally promoting outcrossing (Heinrich, 1975b). Specialized features such as long tubular corollas and leathery calyces are effective means of excluding a great many (but not all) unwanted foragers such as flies, beetles, and small bees from flowers normally pollinated by large hawkmoths, birds, and bats. A dominant characteristic of bird-flowers is that they produce large amounts of nectar, which is sequestered in long, tubular corollas that are red in color (Baker, 1963). Since many insects (including bees) do not have color vision in the red, the flowers are doubly protected from nectar-stealing insects. Bat flowers, on the other hand, generally produce their nectar at night when most bees are not active, and they are strongly odoriferous and often white in color, so as to be conspicuous in the dark (Baker, 1961).

Bumblebee flowers have somewhat more amorphous characteristics than the more distinctive bird and bat flowers. Very few flowers are pollinated by one group of pollinators only, and bumblebees, in particular, visit almost anything. Nevertheless, a few kinds of flowers, because of their morphology, admit bumblebees to their pollen and nectar more readily than other foragers. The flowers specializing in bumblebee pollinators are of great variety, but they have several characteristics in common. Monkshood is a fairly typical bumblebee flower. It occurs only where there are bumblebees, and it is pollinated almost exclusively by them. As one might expect from a plant catering only to relatively large, energy-demanding pollinators, the flowers contain large amounts of nectar. The nectar is found in two modified petals, where it is effectively hidden from most foragers except some specialized bumblebees. It is a curious fact that many of the flowers evolved to be pollinated specifically by bumblebees have hidden nec-

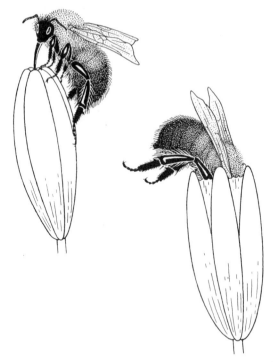

Fig. 11.3 Bumblebees reach the copious nectar of closed gentian blossoms by prying apart the pleated corolla tube and crawling into the base of the flower.

tar or pollen. Apparently this flower morphology prevents other, non-pollinating insects from reaching the food rewards, but the bumblebees are generally versatile enough to harvest from the flowers and pollinate them. It is advantageous to the plant to have individual bees specialize in it, because this promotes intraspecific pollination, a point that will be discussed in more detail in the next chapter. An extreme example of a flower that is difficult, and thus caters to individual foragers, is the closed gentian occurring in the central and northeastern United States. As the name implies, the flower remains permanently closed, yet some individual bumblebees manage to pry apart the petals and reach the nectar and pollen (Fig. 11.3). It is generally well worth their effort, for a single flower of this species may provide up to 45 μl of nectar that is 40 percent sugar, a reward that is an order of magnitude greater than that of most other flowers.

Another group of typical bumblebee flowers, of much different mor-

phology, are the Louseworts (Genus *Pedicularis*, Family Scrophularia-
ceae). Mostly through the work of Lazarus W. Macior (1970; 1973),
who has studied pollination of these plants in great detail, we know
that the approximately 100 species are almost exclusively pollinated
by bumblebees. The plants have evolved an amazing variety of flower
colors and morphologies, but in general the flowers have a two-lipped
tubular corolla.

Bumblebee flowers are generally not distinct from bee flowers in
general, although they tend to be large and have food rewards that are
more ample and hidden. Many bumblebee flowers are bilaterally sym-
metrical, or "zygomorphic." In all of the numerous forms of zygo-
morphy, access to the food rewards is channeled to promote maximum contact between the male and female reproductive organs by a
bee visiting different flowers (see Fig. 11.4). Zygomorphic flowers very
often also require forced entry, so that only relatively intelligent pollinators such as higher or social bees can gain entry, and many of the
behaviorally less versatile foragers are excluded. Many bee flowers also
reflect in the ultraviolet, which is not visible to humans and other vertebrates, but is visible to bees (Daumer, 1958).

Some bumblebee flowers are sufficiently complex morphologically
to exclude many bumblebees too. For example, I have observed that
many inexperienced bumblebees fail to find the nectar rewards in
monkshood (*Aconitum napellus*) flowers. However, those individuals
that do locate the nectar find large amounts, and possibly because of
these ample rewards, they then visit these flowers to the exclusion of
others. A general trend in evolution, therefore, appears to be for greater
pollinator-plant specificity by the exclusion of broad taxonomic group-
ings of potential visitors. Ultimately some individuals of the pollinator
species that specializes in the flowers are also excluded (Heinrich,
1975d). The resulting reproductive isolation is thought to be a prime
mover for speciation in plants.

Although the dense panicles of goldenrod, and the inflorescences of
aster, meadowsweet, and other plants are not bumblebee flowers in
the classical sense,they are often exploited and pollinated by them almost to the exclusion of other pollinators. The individual florets generally contain minute quantities of food reward. But since the florets
are massed, the perching foragers, such as bumblebees, may achieve a
relatively high rate of energy intake by extracting the food from large
numbers of florets in a short time. In addition, as mentioned pre-

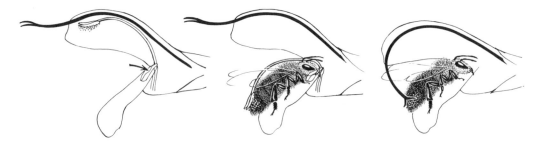

Fig. 11.4 Pollinating mechanism in sage flowers. The first two sketches show the male stage of the flower cycle, with the pistil raised out of contact with the bee. When the insect pushes the shovel-shaped portion of the anthers against movable articulations at the flower base, the two fused anthers move down, dusting the bee's abdomen with pollen. The pollen from one flower is picked up by the pistil of an older flower that has become functionally female by having its pistil recurved to touch the bee's abdomen (far right).

viously, bumblebees can, on occasion, also lower body temperature and concomitantly lower energy expenditure during foraging. The latter option is not available to hoverers, which are excluded from these flowers on energetic grounds. Bumblebees are the major pollinators in some habitats that contain few, and sometimes no, classical bumblebee flowers. Good examples include boreal bogs and mountains. Apparently the number of plant species is low enough in these habitats that it is not necessary to exclude some foragers from the flowers of some plant species to provide pollinator specificity. The bees have to specialize anyway, because they have few options at any one time. The plants have enhanced the effect by staggering their blooming times, rather than by developing complex flower morphologies requiring specialized foraging techniques.

The distance between flowers also affects the energy balance between plants and pollinators. Although this distance is ultimately determined by the population density of the plants, it is altered by the time and duration of flowering. For instance, synchronous blooming of the flowers of a species in a foraging area would minimize the time and energy expenditures of the pollinators flying between plants. Brief blooms of individual flowers, with the plants producing flowers over a long period of time, would maximize the expenditure of the pollinators. If the expenditure of foraging on a plant species is too great relative to the rewards it gives, then the bees will not become conditioned to the flowers and will seek others instead. Synchronous blooming, because it can promote conditioning of the pollinators, should be particularly advantageous in rare plants or in those that produce few rewards. If the rare plant does not bloom synchronously or provide a large food reward then it must evolve some alternate mechanism to be cross-pollinated, or it will have to forgo cross-pollination.

Some orchids—a species-rich group, particularly in the tropics—have evolved a way around the problem of achieving an energy balance with pollinators; they have other flowers in the plant community feed their pollen vectors. Even at very low population densities, the orchids are able to achieve pollination by their extreme precision in pollen transfer and reception. Pollen from a single flower becomes detached in a single mass, the pollinium, which becomes attached to the foraging insect at a specific site on the body and then remains firmly affixed to the insect until it is retrieved at the stigma of another flower of the same species, often one that is far away. These orchid species can thus be pollinated while being very rarely visited. Continuous production of reward, to encourage repeated visits, is not necessary. These plants do not need to maintain an energy balance with their pollinators, who rely mostly on other more common and rewarding flowers for their food energy. Many orchids get by without providing any reward at all; they mimic other flowers that do offer rewards. There are even some orchids that in shape, size, color patterns and scent have flowers that mimic female wasps and bees, including bumblebees. The male bees attempt to copulate with the flowers and thus unwittingly transfer pollen, causing cross-pollination (Kullenberg, 1961). The flowers thus rely on deceit to attract pollinators (details in Chapter 12). More than 10 percent of all flowering plant species are orchids and they are pollinated predominantly by bees.

The more highly evolved orchids typically attract species-specific pollinators. Much of the floral variation among orchids is the result of visual signaling and of floral rearrangement necessary to physically fit specific bees and other pollinators.

One group of bees closely related to bumblebees, the solitary euglossines, or golden bees, which are restricted to the New World tropics, are specialized for pollination of nearly two thousand species of orchids. The golden bees (tribe Euglossini, subfamily Bombinae) consist of six genera: *Eulaema, Euglossa, Euplusia, Eufriesea, Exaerete,* and *Aglae.* (The latter two, like their counterpart the *Psithyrus* in the bum-

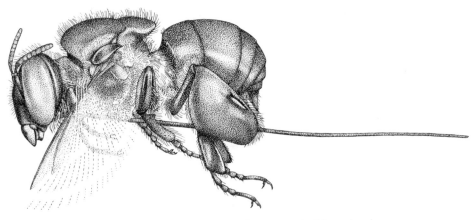

Fig. 11.5 Drawing of a euglossine, or golden, bee, a tropical bee closely re-lated to bumblebees. Note the long tongue trailing behind in flight and the thickened hind tibia.

blebees, parasitize other euglossines.) *Eulaema* and *Euplusia* comprise about sixty species of large hairy bees. *Euglossa* comprises about a hundred species of sparsely haired, medium-sized, brightly metallic blue, green, or golden bees. Some have huge tongues that may be twice the body length (Fig. 11.5). In flight, the tongue is tucked back between the legs and extends far beyond the tip of the abdomen, where it resembles a tail when seen from afar. The males have curious brushlike front tarsi and inflated hind tibia with slit openings and glan-dular interiors.

The males (particularly those of *Eulaema, Euglossa,* and *Euplusia*) scrape specific species of orchid flowers with their tarsal brushes and deposit the flowers' scent material into their inflated hind tibiae. What do they do with this collected perfume? Caloway H. Dodson and his coworkers (1969) have found that the perfume collected by the males of any one species attracts other males of the same species. Small groups of males congregate into groups where they intercept females; similar behavior is well known among some birds, such as the bird of paradise and prairie chicken.

There are many unanswered questions about the bees' mating be-havior. For example, it is not known what attracts the females to the males' aggregation. Does the mere sight of large numbers of males at-tract females, or are the females attracted by the perfume, or both? Females are never attracted to the orchid flowers that yield the scent to the male bees, but it is possible that the males chemically alter the perfume inside their glandular hind tibiae. Why and how the males form groups is also not known. If the females are attracted to a scent that males must laboriously collect and chemically modify, then it would be advantageous for other males to be attracted to the same scent. This would allow them to intercept females without expending time and energy to search for the specific and often rare orchids to collect and make their own perfume.

This sort of behavior is not without precedent. W. Cade (1975) has investigated similar activity in the field cricket, *Gryllus integer*. The males of this species chirp to call females for mating (rather than using a scent, as bees do). The primary cost of this activity is not the energy expended in the chirping; the primary cost is the risk of parasitism that chirping involves. A species of fly, *Euphasiopteryx ochracea*, may cue in on the calling male's song and larviposit on the singer, who ultimately dies when the larvae consume his insides. Obviously, calling for females is hazardous. However, some males, called "satellite males," which remain silent, are attracted to singing males. In these nocturnal aggregations of males, the satellites attempt to intercept and copulate with females that the singer attracts, but they do not pay the attendant cost of singing—parasitism by the fly larvae.

Male bees of a particular euglossine species will collect scent only from flowers of a particular orchid species. Dodson and his co-workers (1969) have investigated this flower-bee specifity by making a chemical analysis of orchid scents, combined with behavioral experiments on golden bees. Most orchid species pollinated by golden bees contained seven to ten, and up to eighteen, distinct chemical compounds. Field testing the isolated compounds, they found that 1,8-cineole was the most general attractant. This compound was found in 60 percent of the orchid samples, and by itself it attracted 70 percent of the bee species. Why, then, were bee species highly specific to certain orchid species? It was found that adding certain compounds resulted in the exclusion of certain bee species (Williams and Dodson, 1972). It appears, therefore, that the orchids achieve flower constancy in specific pollinators by providing specific combinations of attractants and repellents. The scents of orchids are analogous to the flower colors used in signaling by most other plants, except that colors generally attract and repel at the individual rather than the species level.

Although the golden bees are adapted on a one-to-one, species-to-species, basis for the pollination of certain orchids, they rely on other flowers for their food, enhancing their role as pollinators in the South and Central American forest ecosystem. They are all strong flyers with a very high energy expenditure. Like bumblebees and honeybees, they usually die of starvation within an hour or so if confined without food. But their extremely rapid flight allows them to range widely to find flowers with large nectar volumes. They are particularly important pollinators of rare and widely dispersed plants in the understory of the forest. The plants they utilize present few highly rewarding flowers at one time, since they flower over many months rather than blooming synchronously. Dan Janzen (1971) has found that the bees regularly visit series of widely separated plants each day in specific sequences, which he calls "trap lines," to empty newly opened flowers.

Energy economics is a major unifying theme in the mutual interactions of bees and plants. The general principles are the same for bumblebees and for other pollinators, but it is impossible to predict specifics from generalities. The generalities are useful primarily for recognizing and possibly predicting exceptions, for placing observations into rational contexts, and for giving frames of reference (sometimes temporary ones) for further analysis and further questioning into ecology and evolution.

CHAPTER TWELVE

What had that flower to do with being white,
The wayside blue and innocent heal-all?
What brought the kindred spider to that height,
Then steered the white moth thither in the night?
What but design of darkness to apall?—
If design govern a thing so small.

—Robert Frost, *Design*

Ecology and Coevolution

In both physiology and ecology, the primary aim is to understand the interrelationships of component parts. What differs is the whole one takes as the point of reference—the organism or the ecosystem. An organism is more than a collection of cells—it is an integrated system. One studies the functioning of an organism's system by observing the parts, making "models" of how they interrelate, then testing the models by experiments that involve changing the system to observe predicted effects. Similarly, an ecosystem like a forest is more than a collection of trees. By determining the components of the system and arriving at logical constructs of how they interact with each other and with the environment, one produces models or hypotheses. In ecological systems, we are often restricted to observing the results of "experiments" that nature provides, and nature confounds all the variables and seldom provides controls. Furthermore, the time course is different. Removing the heart of a bumblebee would prevent it from flying because the flight muscles could no longer be supplied with sufficient amounts of fuel, and because it might overheat. But removing a vital component from a forest may not yield visible results until decades later. It all adds up to challenging science. In ecological systems, it is seldom possible to remove or ablate a component—such as the pollinators—hence the approach tends to be descriptive and analytical rather than experimental. But change in one part of the system must eventually bring change in another, for the parts have not been assembled at random through evolution. One can ask, then, not only how the system functions at any one time, but also how the system has come into being.

What do we actually know about the details of the coevolution of bees and flowers? Not very much. The fossil record is spotty. Most of our knowledge is by way of inference and extrapolation from existing relationships. We know from the fossil record that the Angiosperm plants (those having enclosed seeds) began to dominate the flora of all latitudes in the latter half of the Cretaceous period, about 100 million years ago (Baker and Hurd, 1968), and we know that a prominent feature of the Angiosperms is the adaptation of their flowers for insect pollination.

The first primitive insects had already appeared in the Upper Carboniferous 300 million years ago, 200 million years before the Angiosperms began their ascendency. Hymenoptera, the insect order to which bees and ants belong, were present 80 million years before that mid-Cretaceous Angiosperm explosion. Indeed, ants, presumably living in societies (there are no solitary ants), had already evolved from their wasplike ancestors. The first primitive insectivorous mammals from which primates and, much later, man evolved, did not appear until about 40 million years later. Bees also evolved from wasps, probably as a result of their switching from predation to pollen foraging to get protein, a change that may have triggered the Angiosperm explosion about 83 to 113 million years ago.

By the Eocene, about 58 million years ago, the Hymenoptera had pretty well differentiated into the groups found today; most of the major groups of bees that we have had already appeared. Examples of many have been preserved in amber (Carpenter, 1976). However, only a few of the 58-million-year-old bees have been tentatively identified as belonging to *Bombus* (Zeuner and Manning, 1976). Less than a half dozen possible bumblebees have been recovered from the subsequent Oligocene and Miocene deposits, but, according to T.D.A. Cockerel, a noted bee taxonomist, only one true *Bombus* fossil has been described. This bee, named *Bombus proavus* by him, closely resembles the European *B. lapidarius* in wing venation (Fig. 12.1). It was found in the Latah formation (Upper Miocene, 20 million years old) near Seattle, Washington, where it was associated with a predominantly temperate fossil flora (Carpenter, 1924).

The fossil record is too incomplete to give us a dynamic picture of coevolution of bees and plants. However, by examining extant animal-

Fig. 12.1 Photograph of the fossil bumblebee *Bombus proavus* Cock. The wing is 15 mm long. (Courtesy of Frank M. Carpenter.)

plant interactions in carefully chosen model systems, we can make deductions about the selective pressures that have been operating to produce adaptation through the process of evolution. Using these systems, we can explore some of the functional interrelationships between bees and flowers, with particular emphasis on bumblebees.

In the last few years pollinator-plant interactions in northern bogs have been extensively studied in research that provides insights into coevolution (Reader, 1975; Small, 1976; Heinrich, 1976a). These bogs contain little of economic interest to man and their assemblages of plants and animals are relatively undisturbed. Bogs are convenient for the study of pollinator-plant relationships, because the species diversity of both plants and pollinators is low. Furthermore, many of the coexisting plants are closely related, belonging to the Ericaceae family (which includes blueberries, azaleas, and cranberries). In most bogs in various parts of Canada (Pojar, 1974) and Maine (Heinrich, 1976c) bumblebees are the most numerous flower visitors, and bees of each species visit the flowers of a succession of different plant species

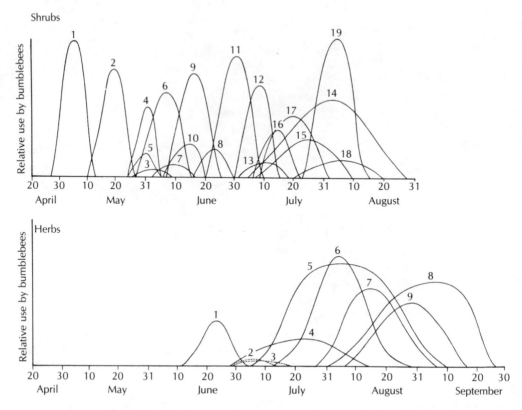

Fig. 12.2   The flowering times, and the relative use by bumblebees, of shrubby and herbaceous plants of Maine bogs. The plants are identified by the numbers shown in Figure 12.3.

throughout the season, from early spring until late fall (Figs. 12.2, 12.3).

The bumblebees may at times rely on only two or three plant species for most of their food, and these plants then play a central role in the ecosystem. There is apparent species-interdependence among the plants, since each of the different plant species provides a link in the temporal chain of food necessary for the bees', and ultimately the plants', survival.

Flower visitation, as such, does not necessarily imply pollination. But the work of R. J. Reader (1977) has shown that flower-visitation by insects is essential for pollination in the eight bog plants in the Family Ericaceae he examined in Ontario. (Fig. 12.4). These studies have provided key evidence linking the bumblebees to the bog ecosystem. Reader emasculated flowers by hand (removing the anthers), enclosed them in mesh bags to exclude all pollinators, and later hand-pollinated them with pollens derived from the same or from other plants, or left them unpollinated. By examining for seed-set, it was shown that all eight species required insect pollination, even though they were all physiologically self-compatible. In all cases, seed-set was abolished by the exclusion of insects from the flowers. Apparently the flowers do not normally self-pollinate; structural features prevent the pollen from contacting the stigma of the same flower. He found, further, that bees were the most common flower visitors, and bumblebees were by far the most common bees. Absolute numbers of specific pollinators, however, did not give an indication of their relative importance for pollination. Bumblebees may have been even more important pollinators than their numbers indicated. Andrenid bees, for example, which decreased in abundance throughout the season, as they do in Maine, were up to six times slower in moving from one flower to another. Assuming they transfer the same amount of pollen between plants (which is highly unlikely), each andrenid bee would be only one-sixth as effective a pollinator as each bumblebee. However, it is not known how much pollen, if any, some of these smaller bees carry between anthers and stigmas of different flowers. Bumblebees, being large and hairy, transfer large amounts of pollen and make ample contacts with the flower stigmata.

Although the central role of bumblebees in bogs is firmly established, there is no reason to suppose that they are the only important pollinators. For example, flowers depleted by bumblebees may be visited by other small insects for the leftovers. And because the remaining food is minimal, these low-energy-demanding animals are forced to visit numerous flowers, although, without the prior activity of the bumblebees, they might need to visit only a few. In addition, it is likely that if bumblebees were absent from the system for several years, then populations of other nectivores would build up and increase their importance as pollinators. Because of fluctuating conditions, it is probably not optimal for a plant to evolve too tight a relationship with any one pollinator. They must hedge their bets and maintain a hierarchy of pollinators. Some apparent inefficiency in the system at any one time may thus be necessary to long-term optimality.

Competition for limiting resources has been a driving force in the

Fig. 12.3 The main plants that bumblebees forage from and pollinate in the bog ecosystem. The first two plates depict the shrubs and vines and the third depicts the herbaceous plants. Both groups are arranged in approximate order

Shrubs and Vines

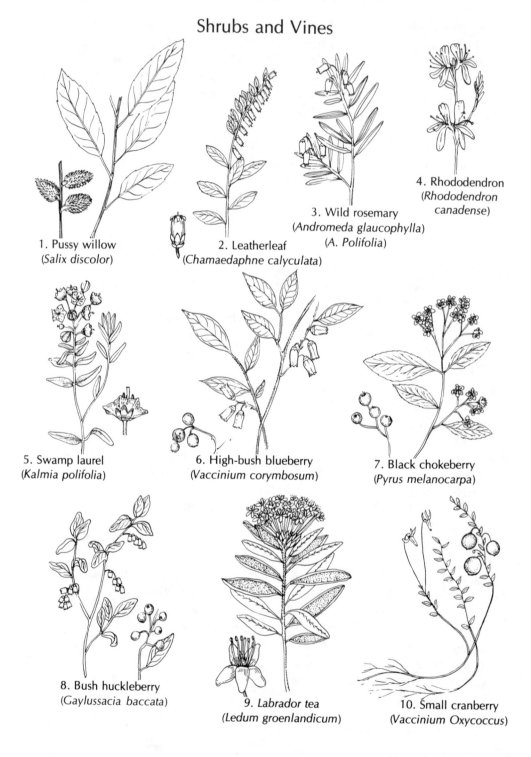

1. Pussy willow
(*Salix discolor*)

2. Leatherleaf
(*Chamaedaphne calyculata*)

3. Wild rosemary
(*Andromeda glaucophylla*)
(*A. Polifolia*)

4. Rhododendron
(*Rhododendron canadense*)

5. Swamp laurel
(*Kalmia polifolia*)

6. High-bush blueberry
(*Vaccinium corymbosum*)

7. Black chokeberry
(*Pyrus melanocarpa*)

8. Bush huckleberry
(*Gaylussacia baccata*)

9. *Labrador tea*
(*Ledum groenlandicum*)

10. Small cranberry
(*Vaccinium Oxycoccus*)

of flowering. In the first plate, all except the willow (*1*) belong to the family Ericaceae. Plants are shown one-third to one-half natural size.

Shrubs and Vines

11. Lambkill
(*Kalmia angustifolia*)

12. Winterberry
(*Ilex verticillata*)

13. Large cranberry
(*Vaccinium macrocarpon*)

14. Field spirea
(*Spiraea latifolia*)

15. Swamp rose
(*Rosa palustris*)

16. Privet andromeda
(*Xolisma ligustrina*)

17. Smooth rose
(*Rosa blanda*)

18. Hardhack
(*Spiraea tomentosa*)

19. Buttonbush
(*Cephalanthus occidentalis*)

Fig. 12.3 *Continued*

Herbs

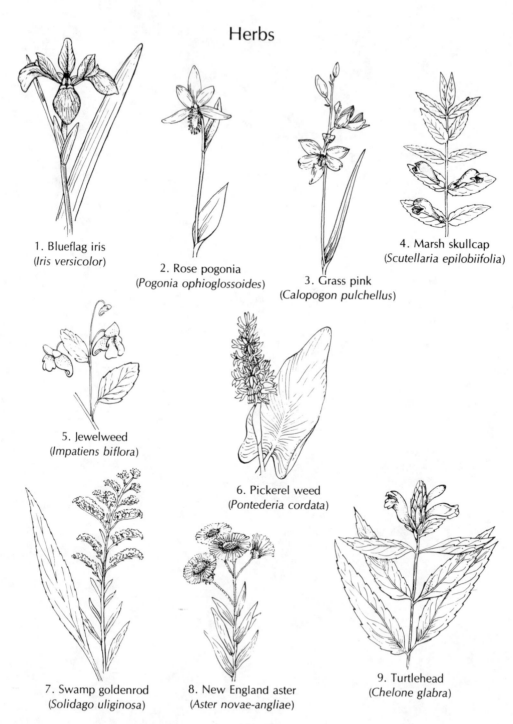

1. Blueflag iris
(*Iris versicolor*)

2. Rose pogonia
(*Pogonia ophioglossoides*)

3. Grass pink
(*Calopogon pulchellus*)

4. Marsh skullcap
(*Scutellaria epilobiifolia*)

5. Jewelweed
(*Impatiens biflora*)

6. Pickerel weed
(*Pontederia cordata*)

7. Swamp goldenrod
(*Solidago uliginosa*)

8. New England aster
(*Aster novae-angliae*)

9. Turtlehead
(*Chelone glabra*)

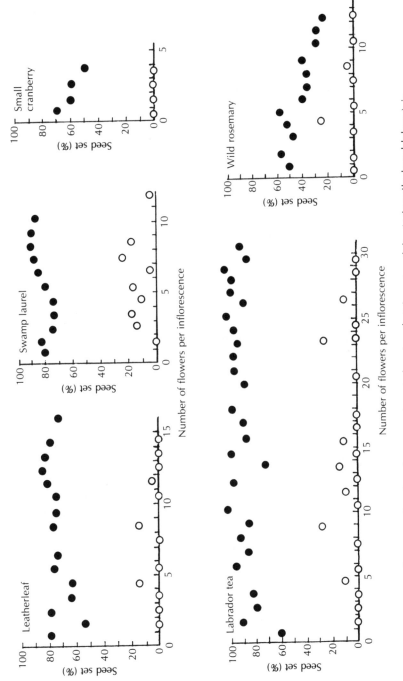

Fig. 12.4 Results of an experiment showing that insect activity (primarily bumblebees) is required to produce seed-set in five species of bog plants (in southern Ontario). The flowers were either left open to pollinators (●) or screened to exclude them (○). The bees pollinated the flowers when they were clustered into small as well as large inflorescences. (From Reader, 1977.)

evolution of numerous physiological, morphological, and behavioral traits of organisms. Competition for food and energy resources has resulted in efficient food detecting, capturing, and processing mechanisms, and competition for mates has resulted in some of the showy and colorful displays of birds, mammals, and some insects. Most plants, being sedentary and bisexual, compete for pollinators that cause fertilization. The evolution of the conspicuous color and scent signals of flowers, as well as various other aspects of their flowering strategies, has undoubtedly been strongly influenced by competition for pollinators. Not only do bumblebees compete for nectar and pollen, but plants simultaneously compete for the services of bees.

If there is competition for nectar the bees are under selective pressure to evolve foraging efficiency. But an unpredictable event, such as an extremely late frost one year, can kill most of the bees, and then the plants must be effective competitors for the remaining pollinators in order to survive. Over short periods of time the two types of competition predominate alternately. But in terms of evolutionary time, competition between plants for pollinators and competition among pollinators for the food rewards of plants operate concurrently.

It is also highly probable that both forms of competition occur simultaneously in an equilibrium condition when resources are relatively, rather than absolutely, limiting. For example, different bees may be competing for the one best species of flower, while the other flower species must compete among themselves for bees.

The relative strengths of the two competitions may vary throughout the season. In Maine, for example, there is, in some springs, an intense competition for the few available species of flowers among the bumblebee queens and the many species of solitary bees. However, in the summer, when many introduced plants, such as clover, are abundantly available in fields, the native plants growing nearby must compete for pollinators. Near the end of the summer, when there are again only a few species of native plants in bloom, the bumblebees, which by then have built up to a high population density, must again compete for nectar and pollen, having to visit plants of very low nectar production.

The relative standing crops of nectar are probably not, by themselves, failproof indicators of the degree of competition. Early in the season, queens must be able to collect food in a relatively short time in order to found colonies successfully. They must do all the work themselves, and an efficient start does much to determine ultimate colony success. Competition means increased foraging times, decreased net energy intake, reduced incubation, and lower rates of worker production. It is interesting that many of the high-nectar flowers in Maine, such as blueberries and rhododendrons, bloom early, while low-nectar flowers, such as asters and goldenrods, bloom late.

The evolved interactions between pollinators and plants sometimes border on the bizarre. There is a curious case of possible double-crossing being perpetrated by the grass pink orchid in Maine bogs, for example (and the strategy is apparently successful in other areas—the plant occurs widely in the eastern United States and Canada). The flowers of this orchid provide no food rewards, but they are a relatively large and conspicuous target, generally of showy pink, against the green sphagnum moss on which they grow. They are visited by searching bees (principally *Augochlora*) that are sampling the available flowers, not yet having established their foraging specialties. Bees generally learn relatively quickly to identify and avoid nonrewarding flowers, but these orchids have several tricks that reduce their chances of being quickly identified by the bees. First, they are unscented. Since scent is used for close-in orientation, the bees are forced to rely heavily on color to identify the flowers (Thien and Marcks, 1972). The second tactic of the flowers, at least in the one bog where I observed many hundreds of individual plants, was variability in color. Most of the plants had bright pink flowers, but there were (at least to the human eye) all gradations from white to purple. Variability within a species increases the difficulty of identification by human taxonomists, and there is no reason to suppose that color variability poses less of a problem to bees. For example, a bumblebee or some other pollinator might learn to avoid the pink flowers, and while trying to visit different flowers, approach the white or the purple, all the while remaining with the same plant species. I watched one *B. fervidus* visit seventeen grass pink orchid flowers in a row before I lost sight of it.

The third tactic in the grass pink's double-cross is that the flowers appear to mimic those of another orchid, the rose pogonia, which does provide nectar. Most of the grass pink flowers are pink and have strong ultraviolet reflectance patterns, like those of the rose pogonia. Furthermore, both species bloom in the same habitat, and at the same time (early in July in Maine). Bumblebees visit the two species alternately. Possibly, those searching bees that visit rewarding rose pogonia flowers stay in the habitat long enough to visit grass pink flowers. The two plants appear to have different sites of pollinia attachment on the bee: grass pink orchids attach the pollinium onto the top of the bee's abdomen, while rose pogonia orchids attach it onto the top of the head.

Most insect-pollinated bog plants utilize the same species of bumblebees, which remain in the habitat throughout the entire season. The food available from the different plants is partitioned in ways that are mutually beneficial. Possibly the most important form of partitioning is that the plants pollinated by bumblebees bloom at different times (Fig. 12.2). In contrast, those species that are wind-pollinated, as well as those plants on the forest floor that are pollinated by solitary bees, bloom relatively synchronously in early spring before leaves close up the forest canopy.

How has the staggering of blooms in bumblebee-pollinated plants arisen? Flowering time is under genetic control, and it can be shown to shift by selection over several generations. What have been the selective pressures in nature? One theory is that staggered flowering arose through competition of plants for pollinators. Plants blooming at the most popular times had to contend with greater competition for pollinators. They could attract flower-constant foragers simply by shifting their blooming period to times when few competing plants were in bloom. Some of the bog shrubs, such as the willows and leatherleaf, bloom before leafing out, when the ponds and quiet streams are still ice-covered and the first queens have just emerged from hibernation. These plants rely on energy stored from the previous year for flowering. Other plants bloom when their leaf buds open, and still others bloom after leafing out. Many herbaceous plants bloom later in the season; they must grow leaves and stems before producing flowers. Plant species with similar flowers often bloom one after another in close succession or with overlap, and the first to bloom has more nectar than those following (Heinrich, 1976a). Possibly the later-blooming species are pollinated by insects that have already been conditioned by the similar, earlier-blooming flowers with superior food rewards, and thus they can get away with producing less nectar themselves.

The coevolution of flowering times is a highly complex issue, and like many other ecological-evolutionary problems it has no single and straight-forward answer. If different plants bloom randomly at different times, and there are enough of them, there are bound to be flowers available at all times, and it is not possible to assign any ultimate cause to the different blooming times. However, a comparison with other plant communities may provide insights. I found, for example, that in disturbed habitats (such as fields and roadsides) colonized by many plants that had been introduced from various parts of the world and could not have coevolved, the flowering times were clumped in mid-summer. By contrast, in bogs, plants not only avoided overlap in time of first blooming; they also had shorter flowering times, in comparison to the plants of the disturbed habitat. These two observations, together with the fact that flowering in bogs is spread throughout the season, suggest that flowering time has, in part, evolved to reduce competition.

No one bog can be considered as a self-contained island with its own unique evolution. Bees are known to fly many miles. Along the Maine coast, for example, they can routinely be seen over the sea, flying between islands. When flowers are absent from one bog or island, they fly to another. Any coevolution that may have occurred, either by shifting of existing flowering times, or by sifting of plant species that persist or die out with time, must be seen from a larger geographical, as well as temporal, perspective. A plant may grow in a particular site for years merely because a migrating bird happened to defecate a seed on suitable soil. It may take decades to determine whether or not a plant is successfully established—that is, is reproducing itself adequately. In summary, random events in space and time prevent evolutionary fine-tuning of the system, and this adds up to difficulties in evaluating possible underlying patterns. Indeed, the patterns themselves may in part be determined by random events.

Besides staggering their blooming times, plants can also enhance their chances of being pollinated by having unique flowers that cater to specific species flower-constant individuals. (Levin and Anderson, 1970; Stiles, 1975) Signaling by flowers to attract specific insects was first described by Christian Conrad Sprengel, a German pastor, at the close of the eighteenth century. His book *Das Entdeckte Geheimnis der Natur im Bau und in der Befruchtung der Blumen* ("The Secret of Nature Revealed in the Form and Fertilization of Flowers") set the stage for all subsequent work in pollination. But his ideas did not gain acceptance in his time because he could not explain the significance of cross-pollination (he turned from the study of plants to that of languages). Charles Darwin, who raised orchids and pigeons and knew that frequent outcrosses increased vigor—even though he could not explain why genetically—read Sprengel's book with great interest and finally brought its merit to light. This set the stage for Hermann Müller, a noted German botanist of the nineteenth century, who studied the flora of the Alps in great detail (Müller, 1873; 1881).

Müller noted that fly-flowers tended to be white and butterfly-flowers red, but bee-flowers exhibited a dazzling array of different colors and shapes. John Lovell, of Waldoboro, Maine, expanded on these ideas in his book, *The Flower and the Bee* (1919), on the bees in the northeastern states. He maintained that the color contrasts between different species of concurrently blooming plants enabled the bees more "easily to remain constant to a single plant species in collecting pollen and nectar," and, "If they were to visit flowers indiscriminately, much pollen would be wasted and much time and effort lost in collecting the nectar."

Lovell's idea that flowers are rendered more conspicuous by color contrasts—not only of flowers with foliage but also of flowers with each other—was logically sound, and it fitted neatly with Müller's observations on the Alpine flora. However, it received little notice because there was at that time a vigorous controversy over whether bees had color vision. (One of the "controlled experiments" used to "prove" that bees were color-blind was the demonstration that they visited variously colored and otherwise identical food dishes in equal numbers regardless of color!) It remained for Karl von Frisch to prove by elegant behavioral experiments that honeybees could indeed distinguish a wide range of colors. This laid the physiological groundwork for the ecological theory that the variety, as such, of the various types of signaling by flowers in a plant community has functional significance in energy economy, both for the plants and for the pollinators.

We now know that the evolution of variety in signaling is probably a general phenomenon. It has been documented in the signaling of coexisting animal, as well as plant, species—firefly flashing, cricket and bird singing, butterfly coloration for sexual recognition, moth coloration to deceive predators, and scent signaling in moths. The many messages, each providing a code for a specific species, is undoubtedly an important aspect of the variety we perceive as beauty in nature. And since those codes are not meant for us, we perceive only a part of this variety (see Fig. 12.5). We do not know how the signals have evolved over the millenia, but we can gain general insights from the study of extant model systems.

Plants of the genus *Clarkia* in the evening primrose family (Onagra-

Fig. 12.5 Some flowers of the daisy family that occur together in Florida. As seen by the human eye, they appear uniformly yellow (top). However, as seen by the insect eye, which can resolve ultraviolet, the flowers show different patterns. The bottom row shows the flowers with ultraviolet absorbing areas darkened. The flower species are, left to right: *Helenium tenuifolium, Bidens mitis, Rudbeckia* sp., *Coreosposis leavenworthii,* and *Heterotheca subaxillaris.* (Adapted from Eisner et al., 1969.)

ceae) provide an example that bears directly on the theory of the evolution of flower variety. The first specimens of the genus were discovered by Meriwether Lewis and William Clark on their historic expedition across the western United States. The plant they found, which is now the type specimen of the genus, was named *Clarkia pulchella. Clarkia pulchella* is wide-ranging in the western United States, but it does not occur in California, where all of the other thirty-six clarkia species occur. All grow in semiarid areas, and few grow outside California.

The various species of clarkia are very similar ecologically and vegetatively. It is common to find six or more species on the same hillside in separate little colonies. The plants appear to be similar—all are erect annuals with thin wiry stems and sparse, long thin leaves. Their flowers are also functionally similar. All are open bee-pollinated flowers with two sets of stamens, one set being larger than the other. The flowers shed pollen first from the long and then from the short stamens, and the stigma of a flower becomes receptive only after the pollen has been shed, thus promoting cross-pollination. In most species, the flowers close at night and reopen in the morning.

The species are clearly differentiated in one important detail: the visual appearance of the flower petals (Fig. 12.6). The petals serve only to signal pollinators, and although they remain erect and brightly colored for up to 12 days if the flower is not pollinated, they change color, or wither, in 5 to 6 days after pollination. What is most significant, however, is that the appearance of the fresh flowers differs strikingly between different species. Petals come in a variety of shapes—they may be round; oval, thin, and bifurcated; thin and trifurcated; spatulate; or broad and lobed. Color varies from lavender and purple to almost white, and the color distribution may be uniform or blotched.

What selective pressures have been operating to produce such variety in clarkia flowers? Do the different flower signals serve to attract specific species of bees? J. W. MacSwain, R. W. Thorp, and Peter H. Raven (1973) spent several years studying clarkias and their associated bees, looking for clues why these very similar plants have diverged so dramatically in their flower signaling. Sampling 119 locations in California, Oregon, and Idaho, they found 102 species of solitary bees on 18 species of clarkia examined. Many of the bees specialized in clarkia, as might be expected, since the pollen of these and other plants in the Onagraceae is shed in unique masses joined by viscid threads that require morphological and behavioral adaptations for efficient harvesting—adaptations that the attendant solitary bees possess. Somewhat surprisingly, however, most of the bees did not specialize in particular species of clarkia. They utilized whichever species was handy, regardless of the type of petals it had. It was concluded that "it would be most difficult to argue that pollination systems had played any role in their [clarkia's] divergence." Yet, nobody has looked at the behavior of individual bees foraging on *Clarkia*. It might very well be that variety has evolved to promote flower-fidelity of individuals rather than populations.

If a bee visited flowers indiscriminately in a mixed colony of several species, it would produce hybrids. In clarkia, hybrids are readily produced, and, like mules, the hybrid between horses and donkeys, these hybrids are sterile. Since these few-flowered plants are annuals and reproduce only once, those receiving pollen from other than their own species would be forestalled from contributing to future generations, as surely as if they had been instantly killed. Although it has not yet been tested directly with clarkia and their bees, it is probable that the more

Fig. 12.6 Some of the various petal shapes of flowers of the genus *Clarkia*.

different concurrently blooming flowers are, the less pollinators are apt to stray and cause mass sterilization in their wake.

The work of Harlan Lewis (1953, 1961) with experimental populations strongly supports the above hypothesis. Starting in 1952, Lewis produced artifically mixed colonies of clarkia in various habitats of southern California by broadcasting mixtures of seeds from eighteen species. He then left the sites unmolested. The seeds of clarkia generally drop from the plant and do not disperse widely. After several years he found that most of the introduced species had disappeared. He concluded that the local extinctions were due, in part, to different adaptations to water stress (Lewis, 1953). The role of pollinators, which was not directly evaluated, could have been an additional factor. In one experiment, Lewis created a mixed population of C. *biloba* and C. *ligulata*. These two species cross readily and produce sterile hybrids. The first year after a mixture of seeds of these two species were sown, the plant spacings were random. But in subsequent years the plants of the two species tended to persist in clumps, and ultimately one species replaced another, to produce single-species colonies. In the mixed colonies, the frequency of hybrids was from 3 to 20 percent. C. *ligulata* was usually eliminated, Lewis observed. Whether the bees played a role is unknown, but clarkia could provide a useful model study system for establishing exactly what effect bees do have on the evolution of flora.

The unravelling of the selective pressures that have produced the great variety of flowers, as well as the similarities between flowers of taxonomically unrelated plants, has occupied many researchers (see Lewis, 1966; Levin, 1972; Levin and Schaal, 1970), but it is a mystery that will never be fully unraveled. We can perceive occasional patterns, and we can garner glimpses of the mechanisms here and there. But we will never know the particulars of specific cases, because we can never recreate the local conditions that shaped the evolution of a particular flower.

The same plant often experiences different selective pressures in different parts of its range—the flowers may be visited by different pollinators in different areas. In addition, after having changed from its ancestral form, a flower may become preadapted to other pollinators with requirements similar to those of its original pollinators (see Fig. 12.7). For example, long-corolla bee flowers can also be visited by hummingbirds and moths, and under the selective pressure of these

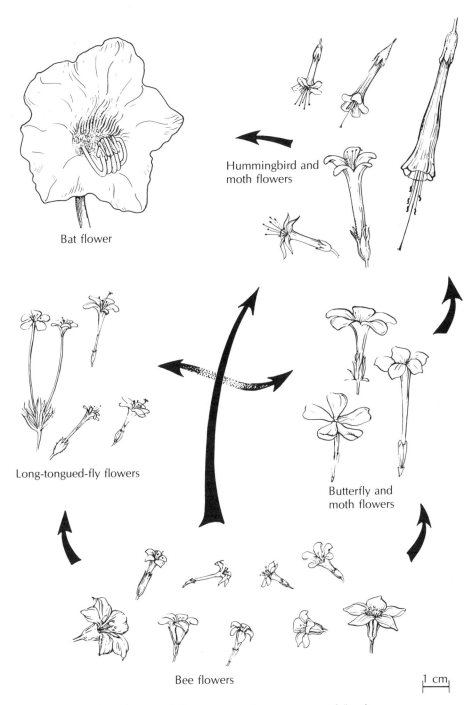

Bat flower

Hummingbird and moth flowers

Long-tongued-fly flowers

Butterfly and moth flowers

Bee flowers

1 cm

Fig. 12.7 Hypothetical route of divergence and convergence of floral morphology under the selective pressure of different pollinators, starting from ancestral bee-pollinated flowers. These flower morphologies are found within the *Phlox* family. (Adapted from V. Grant and K. A. Grant, 1965.)

pollinators they may become further modified until they are also visited by flower-foraging bats. *Cilia splendens,* an annual plant with pink, funnel-shaped flowers, has long slender corolla tubes when growing in the San Gabriel Mountains of California, where it is pollinated by bees and hover flies (*Bombylius*). In the San Bernadino Mountains, where it is pollinated by hummingbirds, it has long, stout corolla tubes. In the desert, where pollinators are less reliably present, *C. splendens* survives by self-pollination (Grant and Grant, 1965).

Presumably, when a particular type of flower is already present in a habitat, then the resistance to the evolution (or introduction) of other flowers of the same type will be greater than if it were not there. If no flowers of a particular type suited for a particular pollinator are present, then there should be less resistance to the evolution or introduction of that flower type, given the availability of a suitable pollinator. The opportunity to exploit that flower type would be open to plants regardless of their taxonomic affinity to other flowers of a similar form. It is only necessary that they be already partially preadapted so that the change required is not too great, and that they be under selective pressure to change. Bees are reliable pollinators and have been utilized by the majority of plant groups, so that many different types of bee flowers have evolved. For example, one highly specialized form of bee flower is vibrated by the bee's buzzing so that it releases pollen onto the insect's venter. This type has evolved in the Solanaceae, Primulaceae, Vacciniaceae, and Leguminaceae, among other plants.

The greater the number of coexisting plant species that require cross-pollination, the greater will be the selective pressure on them to be different from each other, especially among flowers that bloom simultaneously (there is some evidence that consecutively blooming plants can be similar and make use of the same preconditioned foragers). Small wonder then that the most bizarre flowers and pollinating mechanisms are in the tropics with its incredible abundance of species.

Although in most cases coevolution between plants and their pollinators has resulted in mutual benefits, there are exceptions. These exceptions should not be surprising, because in all insect-plant relationships each partner has only its own evolutionary interest in view. Neither partner ever evolves for the sake of the other. It is likely that the majority of flowers are foraged from one or by another insect that takes nectar and pollen without actually doing any pollination. An extreme example is the nectar "stealing" short-tongued bumblebee that bites holes into long corolla tubes of some flowers, thus entirely bypassing the reproductive organs. Many other insects are equally useless as pollinators, even when they enter a flower legitimately, either because they are small and do not contact the flower's reproductive organs, or because they do not reliably carry pollen between different flowers. The cost of nectar thefts is probably negligible to most plants, because they usually get pollinated anyway, except in those rare instances where the thieves remove all of the nectar. In some cases thievery may even help flowers. For example, seed-set in clover fields is sometimes increased when nectar-stealing bumblebees are present, presumably because the reduced amounts of nectar per flower cause the long-tongued bumblebees, the legitimate pollinators, to visit more flowers than they would otherwise.

A great many plant species are self-pollinating, and which insect visits them, and how, is irrelevant. For all intents and purposes, their elaborate flowers—showy signaling devices once meant to attract some erstwhile pollinators—are now useless, and would appear to be an unnecessary drain on the plants' energy resources. However, structures may be much slower to change than immediate selective pressures—a sobering thought to the field biologist who is trying to unravel the adaptive significance of a particular pollination phenomenon in terms of the selective pressures he or she laboriously documents in a summer's field work.

It is clear that many insects take from flowers without giving anything in return, although the cost of this to the plant can be evaluated from many perspectives. Similarly, there are many plants that take advantage of insect pollinators without affording them any benefits. The grass pink orchid, discussed previously, is a prime example. It offers its bee pollinators nothing except the expectation of food. Some other orchids offer insects the expectation of sex. In Europe, the looking-glass orchid, *Ophrys speculum,* which provides no nectar, mimics in scent and appearance a female wasp, *Scolia ciliata.* It attracts *Scolia* males who (presumably) get nothing for their attempts to copulate with numerous flowers, but unwittingly provide the orchids with cross-pollination. Another *Ophrys, O. muscifera,* perpetrates the same deception on a digger wasp, *Gorytes mystaceus,* and in Australia four species of orchids (*Cryptostylis leptochila, C. subalata, C. erecta, C. ovata*) pull the same trick on an Ichneumon wasp, *Lissopimpla semipunctata.* One Ophrys, the bee orchid, *O. apifera,* mimics a bee—possibly a bumblebee. But at the present time this orchid does not succeed in duping bumblebees into pseudocopulation and it is always self-pollinated. We do not know why this particular orchid, which must have been bee-pollinated at one time in the past, is now no longer attracting bees. Possibly the pollinating-bee species died out, or the bees have become smarter and avoid falling victim to the flower's trick. Should the flower, at some future time, again find it greatly advantageous to be cross-pollinated, it will have to perfect its strategy of duping bees.

One of the more interesting ways in which insects get the short end of it is through the trick perpetrated by plants with trap flowers, such as *Arum maculatum*. The flowers attract Chironomid midges by emanating carrion smell (but not providing the genuine article), and, having captured the midges, incarcerate them for several days, dousing them liberally from above with pollen before finally releasing them. The midges are then captured by another flower of the same kind, thus causing cross-pollination.

These examples show how some plants have evolved ways of getting themselves pollinated while providing only the absolute minimum to insects. But, in by far the majority of cases, the plants must pay, and the usual currency is food energy in the form of sugar. Among social bees, the most important pollinators, sugar is the basis of the hive economy. The study of coevolution of plants and pollinators has provided, and will continue to provide, fascinating insights into the role of energy balance in the interactions of organisms in complex ecosystems.

Summary

Bumblebees are able to inhabit cool temperate and arctic regions with short growing seasons in part because they are able to be active at low temperatures, and that ability is a consequence of their remarkable thermoregulatory physiology. The bees can maintain high body and nest temperatures at low air temperatures, but they can do so only by expending prodigious amounts of energy derived from flowers. Since reproductive success is based largely on the amount of energy available for rearing offspring, it is important for the bees to optimize the energy returns per unit of time and energy spent in foraging.

Anatomically, bumblebees are constructed so that heat loss is minimized, as is the energy expended in the shivering necessary to oppose it. Only the thorax, housing the flight muscles, needs to be maintained at a high temperature in order for flight to be possible. Heat loss from the thorax to the surrounding air is minimized by the pile that covers the body and acts as insulation. Conductive heat loss to the abdomen is reduced by the narrow petiole that connects the two body parts and by a large air sac that insulates the anterior portion of the abdomen from the thorax. In addition, a countercurrent heat exchanger retards heat loss to the abdomen while allowing blood to circulate from the thorax.

When food resources from flowers are scarce, the bees economize by cutting down on the energy investment needed to regulate the temperature of the thorax. As a consequence they greatly reduce their speed of locomotion, becoming temporarily flightless while working on some flowers. Their greatly reduced rate of energy use allows them to make an energy profit from the scarce resources, though at a very low rate. Where resources are abundant, however—and the potential for rapid energy profits exists—the bees do not economize in energy expenditure for foraging. Rather, they continue to allocate energy into heat production so that they can maintain a high body temperature and visit flowers at a rapid rate.

A large portion of the energy profit from foraging is used to provide a suitable physical environment for the development of the young. In the nest, the bees convert nectar sugar into heat, which accelerates the developmental rate of the eggs, larvae, and pupae. The adult bees incubate the immatures with their abdomens. The heat is produced by the muscles in the thorax, and the heart physiology is altered so that the countercurrent heat exchange between thorax and abdomen is reduced, allowing the transfer of heated blood from thorax to abdomen. The same mechanism probably operates during flight to remove excess heat from the flight muscles.

All of the raw materials needed for larval growth and to fulfill the bee's energy requirements are derived from flowers. The different flowers require a great variety of specialized foraging skills for efficient exploitation, and individual bees specialize. But the colony as a whole taps a wide range of food resources. Individual bees generally have major and minor specialties, and they switch their foraging investment in accord with the rewards being afforded by the flowers. However, with increasing experience at a particular flower type, a bee becomes increasingly less able to respond to changing rewards.

A bee initially samples a variety of flowers until it encounters one that is satisfying. It then becomes site- and flower-constant. At high bee densities, those flowers that produce large amounts of food are utilized by many bees, and as their rewards are depleted they become equivalent to those offered by initially less-rewarding flowers. Foraging profits are then determined by foraging skill, which is a function of specialization. The bees compete by scramble competition and never by fighting. Fighting bees would be at a competitive disadvantage in making energy profits—the time spent fighting could be spent foraging.

Possibly because of their ability to utilize a large variety of energy resources, coupled with their high energy demands, few bumblebee species coexist in any one area. Different species probably exclude each other by competition for energy resources and nest sites. Those species that do coexist avoid overlap in resource utilization by having different tongue lengths, which affect their ease of access to the nectar in different flowers.

The intricate web of interrelations among bees and flowers is primarily governed by energy needs and payoffs. Throughout their evolution, bees have attempted to reap the maximum rewards from flowers. Flowers, on the other hand, have evolved to supply the least amount of food necessary in order to attract pollinators and to keep them moving from one plant to another for cross-pollination. The bumblebee's energy budget is thus of great significance in the insect's evolutionary game with the flowers. The competition among plants for the pollinators' service, and among the pollinators for the plants' food rewards, decisively shapes the behavior, structure, and physiology of both plants and pollinators. The blooming of flowers in synchronous bursts throughout the growing season, and the variety of flowers at any one time, are some of the evolutionary manifestations of the bees' exploitation of those plants that exploit bees. An investigation of the energetics of the bumblebee reveals that no one way is always best: the bees make compromises that promote long-term success and they vary their strategy to stay in tune with constantly changing conditions. They provide an excellent case study of the operations of evolution and the way a cooperative balance can be established in the ecology of a community.

APPENDIXES

REFERENCES

INDEX

How to Rear Bumblebees

Bumblebees have not been reared on as large a scale as have honeybees, stingless bees, or the leafcutter bees used for alfalfa pollination. Nevertheless, with a little patience and by carefully following a few suggestions and instructions, anyone can keep a bumblebee hive. Some of the instructions are also designed for the professional who may want to cultivate them in the laboratory. Bumblebees readily adapt to captivity, where their entire life cycle can be followed.

Finding and hiving colonies in the field. A good way to begin rearing bumblebees is to collect a colony from the field. Bumblebee colonies are usually not easy to find, but they can be located by watching bees returning to their nest. A bee flying down to the ground (and entering a burrow) is likely to be returning to her colony. (Bees leaving the colony usually fly faster and are less likely to be seen.) Near the beginning of the colony cycle, when there is only one queen and several workers per colony, the chances are remote that an observer will see a bee enter or leave her nest. Later on, when the colonies are populated with many workers, many bees may be entering and leaving per minute, and nests can be more easily located. Nests found in the field can be transferred into almost any kind of a box or hive—a cigarbox with a hole will do. Nests on the surface of the ground can simply be picked up and placed into the hive. Use of a bee veil is highly recommended, particularly when hiving some aggressive surface-nesting species like *Bombus fervidus*. Ground-nesting species can be dug up with a spade: while digging I usually keep an insect net handy and catch bees returning and leaving the nest. The bees are transferred—usually by picking them up by their legs with forceps as they walk up the inside of the net—into a narrow-necked jar. When the nest is reached (usually after digging several feet) the major portion of the colony's population will probably already have been captured and placed into the jar, and the nest can be lifted into the hive. The bees are then dumped from the jar into the nest with one quick motion. (It is generally wise to slap on the hive cover immediately.)

The hive can be placed near a garden to pollinate the plants. It can also be placed indoors with plastic or rubber tubing to serve as an entrance tunnel leading to the outside, or to any other desired feeding area, such as a foraging arena in the laboratory. The nest boxes may be covered with glass so that the bees can be observed. It is often helpful to have a small vestibule outside the nest chamber where the bees can defecate and where a feeder can be located to provide concentrated sugar syrup to the bees while they are becoming established at their new location.

When bees are moved less than a half mile or so, one is apt to lose most of the older worker force; the experienced foragers recognize familiar landmarks upon leaving the nest and return to the original nest location. When colonies are moved far away, however, the bees do not return to the old nest location, but to the new nest instead.

Providing field hives. Bumblebees of many species nest in mouse nests. Mouse nests are often easy to find in early spring where the grass has been matted down by the snow through the winter. They stick up like small mounds. Later, after the grass is tall, the mouse nests—and the bumblebee colonies they may contain—are almost impossible to see. One method to acquire bumblebee colonies is to mark the location of the mouse nests, using tall stakes with flags, in early spring. In late spring, when the grass is high, many of the nests will be occupied by bees.

Lacking a sufficient number of mouse nests to attract the bees, one can set out substitute nest sites, much like setting out bird-houses to attract songbirds. The hives should be set out in early spring just as the first flowers (such as willow) are starting to bloom. This is when the overwintered queens will be searching for nest sites.

The main requirement for a bumblebee hive is a clump of dry fluffy plant fiber in a dark cavity. The bees, depending on the species, will nest in bird-houses, in an old mattress left in the barn, in the pocket of an old coat left hanging in the shed, or in an upside-down flower pot filled with crumpled grass. But for consistent results it is necessary to use more standard techniques. The late G. A. Hobbs and his coworkers (1960) in Canada have been using for many years roughly cubical (160 mm) boxes made of thick plywood painted to exclude moisture. Each box has a removable cover and a 16 mm entrance hole—large enough to admit queens but too small to admit mice. The sturdy boxes prevent skunks from raiding the nests. (However, porcupines some-times chew them to bits, possibly to eat the glue in the plywood.) The boxes should be loosely filled with upholsterer's cotton, and periodi-cally inspected to ensure that the nest material is dry. Boxes with wet nesting material will not be accepted.

Rearing colonies from queens. In the spring one can sometimes cap-ture hundreds of queens in one day. Each queen will ordinarily attempt to start a colony in the field. However, captured queens can, with ef-fort and care, be made to found colonies in confinement (Plowright and Jay, 1966). Chris Plowright of the University of Toronto, who has had more experience than anyone else rearing bumblebees (he has to date raised 46 different species of *Bombus*), prefers to call bumblebee rearing an art rather than a science. Only a small percentage of the queens captured in the field "take" and produce colonies in the labo-ratory, but once a queen starts a colony in the laboratory, the colony may grow more rapidly and to a larger size than in the field. Fifty per-cent success in colony development from captured queens is ex-tremely good.

Before capturing queens, it is necessary to construct and prepare feeder and nest boxes (see Fig. A.1). The plywood feeder box, measur-ing approximately 6″ × 8″ × 4″ high, should contain either a gravity feeder (a pipette will do) or a block of plastic with feeding wells drilled into it. The gravity feeder or feeding well is filled with slightly diluted honey. A nest box, measuring approximately 4″ × 4″ × 2″ high, is placed alongside the feeder box on a platform. Aligned holes drilled into the sides of both boxes allow the bees to travel between them. Only one queen is placed into each bearing setup.

The nest box should be lined with upholsterer's cotton, chip foam, or both. These materials can be obtained from an upholstery supply house or a carpentry supply house, respectively. Some bumblebee

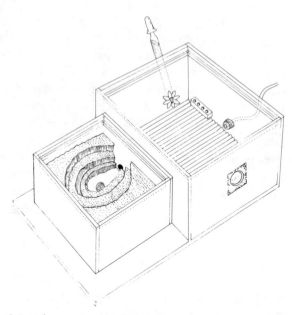

Fig. A.1 Setup for rearing an isolated queen initiating her colony. (See Plowright and Jay, 1966.) Approximate dimensions and other details are given in the text. The nest box (*left*) is drawn to a slightly larger scale than the attached cage (*right*). Note the connecting hole between nest box and cage. The nest box contains a fresh pollen clump surrounded by layers of carpet lining (upholsterers' cotton may also be used). The open tops of the two boxes can be covered with glass, and an additional layer of wood can be added for extra insulation on the nest box. When the colony has become well established, a tube may be provided to allow the bees access to the outside.

species prefer one type of nesting material, and some the other. The primary advantage of chip foam is its ease of handling.

The chip foam, about 5/8″ thick, is layered into the nest box, with enough material cut out of the center to provide a small nest cavity, into which is placed a pollen clump. As the colony grows the nest cavity is enlarged by adding successive rings of carpet lining with more material removed from the centers.

The pollen clump, the food of the presumptive larvae, is crucial for nest initiation. Pollen can be shaken from cattail (*Typhia*) plants in the spring. It can also be taken from honeybees, by erecting a pollen trap at their hive entrance, which strips and collects the pollen loads from the corbiculae of the bees entering the hive. Honeybee pollen purchased from health food stores is generally not fresh enough to provide all of the nutrients needed by the bees. Always freeze the pollen for storage.

The pollen clump is prepared by mixing loose pollen with a few drops of water and a 50 percent honey solution until it is slightly sticky to the touch and has a doughlike consistency. Pollen lumps, about the size of a bean, are made from this dough. A drop of 50 percent honey from an eyedropper is placed onto the nest material to provide a base onto which the pollen clump is lightly pressed. The pollen clump must be replaced every other day until the bee oviposits to make it her brood clump. Finally, moistened filter paper is placed between the top layer of chip foam and the cover to maintain a high nest humidity.

Colony growth and maintenance. After a queen has been freshly captured in the field and placed into her prepared domicile, she must be left undisturbed for at least 24 hours. After 24 hours, the pollen clump should be changed if there is no sign of egg-cell formation. During the egg-cell inspection, the queen can be removed from the nest box (by grasping one of her hind legs with forceps) and placed into the feeder box. The feeder blocks or pipettes should be sterilized periodically. When the feeder box contains feces, it is a sign that the bee is consuming pollen and may soon lay eggs.

As soon as a queen has oviposited into a pollen clump, she begins to incubate it. As she continues to incubate, and the larvae begin to grow, she must be supplied daily with additional small pollen clumps (about 3 mm long, 2 mm wide), which she will incorporate into the brood clump. It is important that she be given no more than she uses, generally no more than one-third the volume of the brood clump that morning. If too little pollen is provided, the queen will uncover the larvae and eject some of them. If there is too much (when excess pollen accumulates), the resulting overfeeding of larvae may stimulate them to develop into queens instead of workers.

Pupation of the first brood will occur approximately 10 to 15 days after the eggs are laid, and new egg batches will be laid on top of the cocoons. The first batch of workers, emerging 16 to 25 days after the first eggs are laid, will initially require plenty of pollen.

By the time the second brood of workers emerges, the colony will probably have to be transferred into a larger receptacle. The bees are not disturbed by light, and a glass cover is acceptable. Throughout colony rearing, the bees can be maintained either in the light or in the dark with no apparent effect on the fecundity of the queen or on the subsequent production of new queens and drones.

Mating and overwintering. The final steps in the complete domestication of bumblebees are to get them to mate and oviposit in captivity. Both can be achieved, although it is more practical to capture new queens each spring. Some bees (*B. vosnesenskii, B. occidentalis*) require large spaces, such as a flight room, for mating, while others will mate even in small nest boxes (*B. rufocinctus, B. appositus*). In order to initiate mating, release about twice as many males as queens at the same time in a flight room that has been provided with a pile of damp peat moss in one corner. Queens that have been inseminated will burrow into the peat moss for hibernation. Mated queens can subsequently be dug out of the peat moss, placed into vials (2″ × 3″) that have been filled with damp peat moss, and then stored at normal refrigeration temperatures (about 5°C). The queens are then left in this condition for their normal hibernation period of 4 to 8 months. Svend N. Holm (1972) of the Royal Veterinary and Agricultural Experimental Farm in Denmark, who has worked extensively in utilizing bumblebees for crop pollination and was one of the first to completely domesticate bumblebees, hibernates queens in mounds of soil in unheated greenhouses, or in plastic containers, with Perlite as bedding, placed into a refrigerator at 4–5°C for 8 to 9 months.

Ernst Horber, an entomologist at Kansas State University, has used bumblebees extensively in his plant breeding program to produce alfalfa and clover varieties resistant to insect and nematode pests. Bumblebees proved to be ideal pollinators in this program because they were the only bees that readily adapted to the small compartment in a greenhouse used for the study.

Horber (1961) kept bumblebee colonies going continuously throughout the year by starting new colonies whenever a new crop of queens hatched. Some of the queens were used to start new colonies, and the rest were stored in a refrigerator in hibernation, to be held in reserve for later use. Colonies of *B. terrestris* with more than 1,500 individuals were produced. By having queens hibernate for various lengths of time, one can have colonies available over a longer period of time. Queens will not found colonies unless they have first spent some time in hibernation, but they can be refrigerated for a longer-than-normal period and still have good (25 percent will survive) colony success. Six months of refrigeration will result in about a 50 percent mortality; ten months will produce about a 75 percent mortality, but colony success will be good (Plowright, personal communication). Such mortalities are probably not excessive; they are greatly exceeded in the field, where one colony may produce hundreds of new queens, and on the average only one produces a successful colony.

The Bumblebees of North America

Bumblebees have long been of interest to biologists, and much litera-
ture is available on their taxonomy. Paul Hurd in his (1978) review of
the superfamily Apoidea lists sixty-four references on bumblebee taxonomy alone. Most of this literature is for specialists. The amateur,
who might like to know more about a bee seen on a blossom, will
probably not get very far in identifying it unless he or she has a net and
killing jar, a microscope, an entomological library, and training in the
use of entomological identification keys.

There are no short-cuts to proper identification. To make a positive
identification of some of the more difficult species, it is usually necessary to enlist the aid of a professional entomologist working on the taxonomy of bumblebees. Part of the problem is that the taxonomy of the
bumblebees of North America is still fairly fluid. It is almost certain that
revisions, at least at the species level, will appear within the next few
years. Different experts still have different opinions as to whether some
bumblebees represent different species or just subspecies.

The color plates (see end of book) and this brief discussion will,
however, provide a superficial aid—reflecting the current taxonomic
consensus—to the identification of the fifty species and subspecies of
bumblebees that Hurd has recognized for the continental United
States, including Alaska and Canada. I have included, additionally, a
few of the Mexican and Central American species. The two color
plates will not provide an infallible guide to all species. Some species,
like *B. crotchii*, *B. flavifrons*, *B. rufocinctus* and *B. sandersoni*, are

polymorphic. The extreme color variations in these species are included on the plates. Other species converge in color patterns. For example, *B. fervidus, B. borealis* and *B. appositus* have similar color patterns, but *borealis* and *appositus,* in contrast to *fervidus,* are tawny yellow, and *appositus* has a paler head and front of the thorax. *B. californicus, B. vandykei* and *B. vosnesenskii* look superficially identical, but *californicus* has a yellow fourth abdominal segment and is black on the face, while the other two species are yellow on the face and on the posterior three-fourths of the fourth abdominal segment. I have included *Psythirus crawfordi,* although in fact it may be merely a clinal variant of *P. insularis.* Three dubious species rarely found in collections—*B. cockerelli, B. pleuralis,* and *B. strenuus*—are not included. I have also not included *B. vagans bolsteri,* a form of *B. vagans* found on the island of Newfoundland.

Some of the bees on the two plates are more commonly known by other names. In cases of recent taxonomic revision I have noted the older name in brackets. All of the North American bumblebee species are included, and even though the color plates are necessarily inadequate for complete identification, it should be possible by a process of elimination to arrive at a narrow range of possible species for any individual bee. But for serious, precise identification it is necessary to examine, with the aid of a microscope, the morphology of one or more of the following features: sting sheaths, copulatory apparatus, mandibles, eye placement, head width, and wing venation. For morphological data on North American bumblebees the works of Milliron (1973), Mitchell (1962), Thorp (1978), Franklin (1912–1913), Frison (1927), Plath (1934), Stephen (1957) and Richards (1968) should be consulted. The works of Franklin and Milliron in particular provide comprehensive information of the bees of the Western Hemisphere. Milliron includes useful range maps of the different species. Thorp's recent monograph gives many details about the bees of California and the western United States.

There are several insects that superficially resemble *Bombus* bumblebees. These include carpenter bees (*Xylocopa* spp.) and some flies that mimic bumblebees and thereby gain protection from predation. The carpenter bee has sparse, short hair. It lacks corbiculae, and the bottom of its eyes nearly touches the base of its mandibles. The bumblebee-mimicking flies have very short, clubbed antennae, and only two, rather than four, wings. (The two wings on each side in the bee are normally hooked together so that they superficially appear as one.)

Psithyrus bumblebees also lack corbiculae, and they are less furry than *Bombus*. Their shiny black exoskeleton is visible on the abdomen and thorax, as it is in carpenter bees. But unlike carpenter bees, *Psithyrus* bumblebees have relatively long individual hairs, and their abdomens are round, pointed, and heavily armored—they do not compress readily when squeezed between the fingers. (It should be noted, however, that the stinger is highly robust.)

Bombus bumblebee queens and workers (all females) are morphologically not much different except for size, but the queens of some species tend to be more bright, conspicuous, and easier to identify than workers. The bumblebees on the two plates were drawn from females. Males (drones) generally have longer antennae and larger eyes than females. They also lack a stinger. The abdomen of drones is tipped with a complex copulatory apparatus, which can be seen, however, only after capturing the bee and squeezing its abdomen. The shape of this apparatus varies from species to species, thus ensuring precise fit with females of the same species during copulation. It also provides a reliable guide to species identification. Color patterns of males are generally similar to those of females, but not as distinct.

All of the bumblebees that are encountered in the early spring (queens) and most of those seen in summer (workers) and early fall are females. The males and new queens are not found in the field until late summer and fall. Males tend to be much more conspicuous than new queens in the fall, because the new queens go into hibernating quarters soon after mating, while the males continue to fly until they die.

GENERA, SUBGENERA, AND RANGES OF NORTH AMERICAN BUMBLEBEES

Adapted from Hurd, 1978. The numbers in parentheses are keyed to the color plates.

Bombus

[A]   subgenus *Bombus* Latreille

B. affinis Cresson—Que. and Ont. south to Ga., west to S.Dak. and N.Dak. (13)

B. lucorum lucorum (L.) [= *B. moderatus*]—Holarctic; Alaska south to parts of southern B.C. and Alta., east through Yukon and N.W.T. (15)

B. terricola terricola Kirby—N.S. to Fla., west to B.C., Mont., and S.Dak. (20)

B. terricola occidentalis Greene [= *B. occidentalis*]—Alaska south to northern Calif., Nev., Ariz., N.Mex., and S.Dak. (25)

[B]   subgenus *Fraternobombus* Skorikov

B. fraternus (Smith)—N.J. to Fla., west to N.Dak., S.Dak., Nebr., Colo., N.Mex. (1)

[C]   subgenus *Bombias* Robertson

B. nevadensis auricomus (Robertson)—Ont. to Fla., west to Tex., Okla., Colo., Wyo., Mont., and southern Canada (Sask., Alta., B.C.). (4)

B. nevadensis nevadensis Cresson—Alaska south to Calif., Ariz., N.Mex., east to Wis.; Mexico. (5)

[D]   subgenus *Separatobombus* Frison

B. griseocollis (Degeer)—Que. south to Fla., west to B.C., Wash., Oreg., and norther Calif. (9)

B. morrisoni Cresson—B.C. to Calif., east to S.Dak., Nebr., Colo., and N.Mex. (6)

[E]   subgenus *Crotchibombus* Franklin

B. crotchii Cresson—Calif., Mexico (Baja Calif.). (2–3)

[F]   subgenus *Cullumanobombus* Vogt

B. rufocinctus Cresson—N.S., N.B., and Que. west to B.C., south to Calif., Ariz., N.Mex., Kans., Minn., Ill., Mich., N.Y., Vt., and Maine; Mexico. (26–28)

[G]   subgenus *Pyrobombus* Dalla Torre

B. bifarius bifarius Cresson—B.C., Oreg. (Steens Mts.), Calif. (Sierra Nevada), Idaho, Utah, Colo. (30)

B. bifarius nearcticus Handlirsch—Alaska and Yukon south to Calif. (Sierra Nevada) and Utah. (29)

B. bimaculatus Cresson—Ont. and Maine south to Fla., west to Ill., Kans., Okla., and Miss. (11)

B. caliginosus (Frison)—Wash., Oreg., and Calif. (coastal areas). (21)

B. centralis Cresson—B.C. and Alta. south to Calif., Ariz., and N.Mex. (44)

B. cockerelli Franklin—N.Mex., Utah.

B. edwardsii Cresson—Oreg., Calif., Nev. (34)

B. flavifrons dimidiatus Ashmead.—Southern B.C. to Calif. (41–42)

B. flavifrons flavifrons Cresson—Alaska south to Calif., Idaho, and Utah. (43)

B. frigidus Smith—Alaska and N.W.T. south to high elevations in Colo. (18)

B. huntii Greene—B.C. and Alta. south to Calif., Nev., Utah, and N.Mex. (40)

B. impatiens Cresson—Ont. and Maine south to Fla., west to Mich., Ill., Kans., and Miss. (12)

B. melanopygus Nylander—Alaska south to northern Calif., Idaho, and Colo. (35)

B. mixtus Cresson—Alaska south to Calif., Idaho, and Colo. (46)

B. perplexus Cresson—Alaska to Maine, south to Wis., Ill., and Fla.; Alta. (10)

B. pleuralis Nylander—Rocky Mt. states, B.C., N.W.T., Yukon, Alaska.

B. sandersoni Franklin—Ont. to Newf., south to Tenn. and N.C. (16–17)

B. sitkensis Nylander—Alaska south to Calif., Idaho, Mont., and Wyo. (45)

B. sylvicola Kirby—Alaska east to Newf., south on the principal cordilla of western U.S. (Cascade, Sierra Nevada, Great Basin, and Rocky Mts.) to Calif., Nev., Utah, and N.Mex. (39)

B. ternarius Say—Yukon east to N.S., south to Ga., Mich., Kansas, Mont., and B.C. (19)

B. vagans vagans Smith—B.C. east to N.S., south to Ga., Tenn., S.Dak., Mont., Idaho, Wash. (14)

B. vagans bolsteri Franklin—Newf.

B. vandykei (Frison)—Wash. to southern Calif. (22)

B. vosnesenskii Radoszkowski—B.C. south to Calif., Nev., Mexico (Baja Calif.). (23)

[H]   subgenus *Alpinobombus* Skonikov

B. balteatus Dahlhom—Holarctic; Arctic Alaska and Canada, south on principal cordilla of western N.Amer. to Calif. (Sierra Nevada and White Mts.) and N.Mex. (Truchas Peak). (31)

B. hyperboreus Schönherr—Holarctic (circumpolar); Arctic Alaska and Canada (Yukon and N.W.T.), Greenland. (32)

B. polaris polaris Curtis—Holartic (circumpolar); Arctic Alaska, Canada, Greenland, and parts of Arctic Eurasia. (33)

B. strenuus Cresson—Alaska, Yukon, and N.W.T. south to B.C.

[I]   subgenus *Subterraneobombus* Vogt

B. appositus Cresson—B.C. east to Sask., south to N.Mex., Ariz., and Calif. (Cascade Mts. and Sierra Nevada). (38)

B. borealis Kirby—Southern Canada from N.S. to Alta., and northern U.S. from Maine to N.J., west to N.Dak. and S.Dak. (36)

B. californicus Smith—B.C. and Alta. south to Calif., Ariz., and N.Mex.; Mexico (Baja Calif. and Sonora). (24)

[J]   subgenus *Fervidobombus* Skorikov

B. fervidus fervidus Fabricius—Que. and N.B. south to Ga., west to B.C., Wash., Oreg., and Calif; Mexico (Chihuahua). (37)

B. pennsylvanicus pennsylvanicus (Degeer) [= *B. americanorum*]—Que. and Ont. south to Fla., west to Minn., S.Dak., Nebr., Colo., and N.Mex.; Mexico and possibly Central America. (7)

B. pennsylvanicus sonorus Say [= *B. sonorus*]—Tex. west to Calif., Mexico. (8)

Psythirus

P. ashtoni (Cresson)—P.E.I. west to Sask., south to N.Dak., Minn., Wis., Mich., Ohio, W.Va., and Va. (59)

P. citrinus (Smith)—P.E.I. and N.B. south to Fla. and Ala., west to S.Dak. and N.Dak. (60)

P. fernaldae Franklin—Alaska and Canada south to N.C. and Tenn. in eastern U.S. and Colo. and Calif. in west. (54)

P. insularis (Smith)—Canada south to Calif., Ariz., N.Mex., Nebr., N.Y., N.H., ?Alaska. (56)

P. suckleyi (Greene)—Alaska south to Calif., Utah, and Colo. (58)

P. variabilis (Cresson)—Ohio south to Fla., west to N.Dak., S.Dak., Nebr., Kan., Okla., Tenn., and N.Mex.; Mexico. (55)

References

Much has been written about bumblebees. In 1912 F. W. L. Sladen published a book on bumblebees (or humble-bees, as they were then called) entitled *The Humble-bee, its Life History and How to Domesticate It.* This book dealt exclusively with British bumblebees. An update of bumblebee behavior and social life was given in *Bumblebees,* by John B. Free and Colin G. Butler, in 1959. This book also concentrated on the biology of the British species. The most recent update has been by D. V. Alford in his 1975 book, *Bumblebees,* again with emphasis on British bees. The only book concentrating on the American bumblebees is Otto E. Plath's *Bumblebees and Their Ways,* published in 1934. Plath (father of the poet Sylvia Plath) describes the habits, life cycles, nests, distributions, and favorite flowers of the thirteen common Eastern *Bombus* species and of four *Psithyrus* species. T. H. Frison and Gordon Hobbs provided much pioneering information on the North American species, published in numerous papers. A detailed comparison of the biology of bumblebees and other bees is found in the 1974 book by Charles D. Michener, *The Social Behavior of the Bees,* as well as in E. O. Wilson's *The Insect Societies.* The economic importance of bees in relation to crop pollination is discussed by J. B. Free in *Insect Pollination of Crops,* published in 1970. General aspects of pollination are presented in the recent books by Proctor and Yeo (1972), and by Faegri and van der Pijl (1971).

Alford, D. V. 1975. *Bumblebees*. London: Davis-Poynter.

Allen, T., S. Cameron, R. McGinley, and B. Heinrich. 1978. The role of workers and new queens in the ergonomics of a bumblebee colony (Hymenoptera:Apoidea). *J. Kans. Entomol. Soc.* 51(3):329–342.

Baker, H. G. 1961. The adaptation of flowering plants to nocturnal and crepuscular pollinators. *Q. Rev. Biol.* 36:64–73.

Baker, H. G. 1963. Evolutionary mechanisms in pollination biology. *Science* 139:877–883.

Baker, H. G., and P. D. Hurd, Jr. 1968. Intrafloral ecology. *Ann. Rev. Entomol.* 13:385–414.

Bawa, K. S. 1974. Breeding systems of tree species of a lowland tropical community. *Evolution* 28:85–92.

Best, L. S. and Bierzychudek, P. 1978. Coevolution of foxglove (*Digitalis purpurea*) and its pollinators:a test of optimal foraging theory. (In preparation.)

Beutler, R. 1951. Time and distance in the life of the foraging bee. *Bee World* 32:25–27.

Bohart, G. E. 1972. Management of wild bees for the pollination of crops. *Ann. Rev. Entomol.* 17:287–312.

Brian, A. D. 1950. The pollen collected by bumblebees. *J. Anim. Ecol.* 20:191–94.

Brian, A. D. 1952. Division of labor and foraging in *Bombus agrorum* Fabricius. *J. Anim. Ecol.* 21:223–240.

Brian, A. D. 1957. Differences in the flowers visited by four species of bumblebees and their causes. *J. Anim. Ecol.* 26:71–98.

Bruggemann, P. F. 1958. Insect environment of the high Arctic. In *Proceedings of the Tenth International Congress of Entomology*, vol. 1, pp. 695–702.

Cade, W. 1975. Acoustically oriented parasitoids: fly phonotaxis to cricket song. *Science* 190:1312–1313.

Carpenter, F. M. 1924. Insects from the Miocene (Latah) of Washington. *Ann. Entomol. Soc. Am.* 24:307–309.

Carpenter, F. M. 1976. Geological history and evolution of the insects. In *Proceedings of the Fifteenth International Congress of Entomology*, Washington, pp. 63–70.

Dansereau, P., and F. Segadas-Vianna. 1952. Ecological Study of the peat bogs of Eastern North America. I. Structure and evolution of vegetation. *Can. J. Bot.* 30:490–520.

Daumer, K. 1958. Blumenfarben wie sie die Bienen sehen. *Z. Vergl. Physiol.* 41:49–110.

Dodson, C. H., R. L. Dressler, H. G. Hills, R. M. Adams, and N. H. Williams. 1969. Biologically active compounds in orchid fragrances. *Science* 164:1243–1249.

Eisner, T., R. E. Silberglied, D. Aneshansley, J. E. Carrel, and H. C. Howland. 1969. Ultraviolet video-viewing: the television camera as an insect eye. *Science* 166:1172–1174.

Faegri, K., and L. van der Pijl. 1971. *The principles of pollination ecology.* 2d ed. Oxford: Pergamon.

Frankie, G. W., P. A. Opler, and K. S. Bawa. 1976. Foraging behavior of solitary bees: implications for outcrossing of a neotropical forest species. *J. Ecol.* 64:1049–1057.

Franklin, J. J. 1912–1913. The Bombidae of the New World. *Trans. Amer. Entomol. Soc.* 38:177–486.

Free, J. B. 1958. Attempts to condition bees to visit selected crops. *Bee World* 39:221–230.

Free, J. B. 1960. The behavior of the honeybees visiting flowers of fruit trees. *J. Anim. Ecol.* 29:385–395.

Free, J. B. 1963. The flower constancy of honeybees. *J. Anim. Ecol.* 32:119–131.

Free, J. B. 1968. Dandelion as a competitor to fruit trees for bee visits. *J. Appl. Ecol.* 5:169–178.

Free, J. B. 1970. The flower constancy of bumblebees. *J. Anim. Ecol.* 39:395–402.

Free, J. B. 1970 *Insect pollination of crops.* New York: Academic Press.

Free, J. B., and C. G. Butler. 1959. *Bumblebees.* London: Collins.

Frisch, K. von 1967. *The dance language and orientation of bees.* Cambridge: Belknap Press of Harvard University Press.

Frison, T. H. 1972. A contribution to out knowledge of the relationships of the Bremidae of America north of Mexico. *Trans. Amer. Entomol. Soc.* 53:51–78.

Gabritschevsky, E. 1926. Convergence of coloration between American pilose flies and bumblebees (*Bombus*). *Biol. Bull.* 51: 269–287.

Grant, V., and K. A. Grant. 1965. *Flower pollination in the* Phlox *family.* New York: Columbia University Press.

Hamilton, W. J., and K. E. F. Watt. 1970. Refuging. *Ann. Rev. Ecol. Syst.* 1:263–284.

Hasselrot, T. B. 1960. Studies on Swedish bumblebees (Genus *Bombus* Latr.): their domestication and biology. *Opusc. Entomol. Suppl.*17:1–192.

Heinrich, B. 1971. Temperature regulation in the sphinx moth, *Manduca sexta. J. Exp. Biol.* 54:141–152.

Heinrich, B. 1972a. Temperature regulation in the bumblebee, *Bombus vagans*: a field study. *Science* 175:183–187.

Heinrich, B. 1972b. Energetics of temperature regulation and foraging in a bumblebee, *Bombus terricola* Kirby. *J. Comp. Physiol.* 77: 49–64.

Heinrich, B. 1972c. Patterns of endothermy in bumblebee queens, drones and workers. *J. Comp. Physiol.* 77:65–79.

Heinrich, B. 1972d. Physiology of brood incubation in the bumblebee queen, *Bombus vosnesenskii. Nature* 239:223–225.

Heinrich, B. 1973. The energetics of the bumblebee. *Sci. Am.* 228:96–102.

Heinrich, B. 1974a. Thermoregulation in bumblebees. I. Brood incubation by *Bombus vosnesenskii* queens. *J. Comp. Physiol.* 88: 129–140.

Heinrich, B. 1974b. Pheromone induced brooding behavior in *Bombus vosnesenskii* and *B. edwardsii* (Hymenoptera: Bombidae). *J. Kansas Entomol. Soc.* 47:396–404.

Heinrich, B. 1974c. Thermoregulation in endothermic insects. *Science* 185:747–755.

Heinrich, B. 1975a. Thermoregulation in bumblebees. II. Energetics of warm-up and free flight. *J. Comp. Physiol.* 96:155–166.

Heinrich, B. 1975b. Energetics of pollination. *Ann. Rev. Ecol. Syst.* 6:139–170.

Heinrich, B. 1975c. The role of energetics in bumblebee-flower-interactions. In *Coevolution of animals and plants,* ed. L. E. Gilbert and P. H. Raven, pp. 141–158. Austin and London: University of Texas Press.

Heinrich, B. 1975d. Bee flowers: a hypothesis on flower variety and blooming times. *Evolution* 29:325–334.

Heinrich, B. 1976a. Flowering phenologies: bog, woodland, and disturbed habitats. *Ecology* 57:890–899.

Heinrich, B. 1976b. Foraging specializations of individual bumblebees. *Ecol. Monogr.* 46:105–128.

Heinrich, B. 1976c. Resource partitioning among some eusocial insects: bumblebees. *Ecology* 57:874–899.

Heinrich, B. 1976d. Bumblebee foraging and the economics of sociality. *Am. Sci.* 64:384–395.

Heinrich, B. 1976e. Heat exchange in relation to blood flow between thorax and abdomen in bumblebees. *J. Exp. Biol.* 64:561en585.

Heinrich, B. 1977a. The exercise physiology of the bumblebee. *Am. Sci.* 65:455–465.

Heinrich, B. 1977b. Why have some animals evolved to regulate a high body temperature? *Am. Natur.* III:623–640.

Heinrich, B. 1978. "Majoring" and "Minoring" in bumblebees: an experimental analysis. *Ecology* (in press).

Heinrich, B., and G. A. Bartholomew. 1972. Temperature control in flying moths. *Sci. Am.* 226:71–77.

Heinrich, B., and T. M. Casey. 1978. Heat transfer in dragonflies: 'fliers' and 'perchers.' *J. Exp. Biol.* 74:17–36.

Heinrich, B., and A. Kammer. 1973. Activation of the fibrillar muscles in the bumblebee during warm-up, stabilization of thoracic temperature and flight. *J. Exp. Biol.* 58:677–688.

Heinrich, B., P. Mudge, and P. Deringis. 1977. A laboratory analysis of flower constancy in foraging bumblebees: *Bombus ternarius* and *B. terricola. Behav. Ecol. Sociobiol.* 2:247–266.

Heinrich, B., and P. H. Raven. 1972. Energetics and pollination ecology. *Science* 176:597–602.

Hobbs, G. A. 1962. Further studies on food-gathering behavior of bumblebees (Hymenoptera: Apidae). *Can. Entomol.* 94:538–541.

Hobbs, G. A., J. F. Virostek, and W. O. Nummi. 1960 Establishment of *Bombus* spp. (Hymenoptera: Apidae) in artificial domiciles in southern Alberta. *Can. Entomol.* 92:868–872.

Holm, S. N. 1966. The utilization and management of bumblebees for red clover and alfalfa seed production. *Ann. Rev. Entomol.* 11:155–182.

Holm, S. N. 1972. Weight and life length of hibernating bumble bee queens (Hymenoptera: Bombidae) under controlled conditions. *Entomol. Scand.* 3:313–320.

Horber, E. 1961. Beitrag zur Domestikation der Hummeln: Vierteljahrsschrift der Naturf. *Gesellschaft Zürich* 106:425–447.

Hurd, P. D., Jr. 1978. Superfamily Apoidea. In *Catalog of Hymenop-*

tera in America north of Mexico, vol. 2, ed. K. V. Krombein, P. D. Hurd, Jr., D. R. Smith, and B. D. Burks. Washington, D.C.: Smithsonian Press. (In press.)

Hurd, P. D., E. G. Linsley, and T. W. Whitaker. 1971. Squash and gourd bees (*Peponapis, Xenoglossa*) and the origin of the cultivated *Cucurbita. Evolution* 25:218–234.

Husband, R. W. 1977. Observations on colony size in bumblebees (*Bombus* spp.). *Great Lakes Entomol.* 10:83–85.

Ikeda, K., and E. G. Boettiger. 1965. Studies on the flight mechanism of insects. *J. Insect Physiol.* 11:779–789.

Inouye, D. W. 1977. Species structure of bumblebee communities in North America and Europe. In *The role of arthropods in forest ecosystems,* ed. W. J. Mattson, pp. 35–40. New York: Springer-Verlag.

Ishay, J., and F. Ruttner. 1971. Thermoregulation in Hornissennest. *Z. vergl. Physiol.* 72:423–434.

Janzen, D. H. 1971. Euglossine bees as long-distance pollinators of tropical plants. *Science* 171:203–205.

Johnson, L. K., and S. P. Hubbel. 1974. Aggression and competition among stingless bees: field studies. *Ecology* 55:120–127.

Jones, C. E., and S. L. Buchman. 1974. Ultraviolet floral patterns as functional orientation cues in hymenopterous pollination systems. *Anim. Behav.* 22:481–485.

Kammer, A. E., and B. Heinrich. 1972. Neutral control of bumblebee fibrillar muscles during shivering. *J. Comp. Physiol.* 78:337–345.

Kammer, A., and B. Heinrich. 1974. Metabolic rates related to muscle activity in bumblebees. *J. Exp. Biol.* 61:219–227.

Kevan, P. G. 1975. Forest application of the insecticide Fenitrothion and its effect on wild bee pollinators (Hymenoptera: Apoidea) of Lowbush blueberry (*Vaccinium* spp.) in southern New Brunswick. *Canada. Biol. Conserv.* 7:302–309.

Knee, W. J., and J. T. Medler. 1965. The seasonal size increase of bumblebee workers (Hymenoptera: *Bombus*). *Can. Entomol.* 97:1149–1155.

Kullenberg, B. 1961. Studies on *Ophrys* L. pollination. *Zool. Bidr. Uppsala* 34:1–340.

Levin, D. A. 1972. Low frequency disadvantage in the exploitation of pollinators by corolla variants in *Phlox. Am. Natur.* 106:453–460.

Levin, D. A., and W. W. Anderson. 1970. Competition for pollinators between simultaneously flowering species. *Am. Natur.* 104:455–467.

Levin, D. A., and B. A. Schaal. 1970. Corolla color as an inhibitor of interspecific hybridization in *Phlox. Am. Natur.* 104:273–283.

Lewis, H. 1953. The mechanism of evolution in the genus *Clarkia. Evolution* 7:1–20.

Lewis, H. 1961. Experimental sympatric populations of *Clarkia. Am. Natur.* 95:155–168.

Lewis, H. 1966. Speciation in flowering plants. *Science* 152:167–172.

Linsley, E. G. 1958. The ecology of solitary bees. *Hilgardia* 27:543–599.

Lovell, J. H. 1919. *The flower and the bee.* London: Constable.

Macevicz, S., and G. Oster. 1976. Modelling social insect populations II: Optimal reproductive strategies in annual eusocial insect colonies. *Behav. Ecol. Sociobiol.* 1:265–282.

McGregor, S. E., S. M. Alcorn, E. B. Kurtz, Jr., and G. D. Butler, Jr. 1959. Bee visits to Saguaro flowers. *J. Econ. Entomol.* 52:1002–1004.

Macior, L. W. 1970. The pollination ecology of *Pedicularis* in Colorado. *Am. J. Bot.* 57:716–728.

Macior, L. W. 1973. The pollination ecology of *Pedicularis* on Mount Rainier. *Am. J. Bot.* 60:863–871.

MacSwain, J. W., P. H. Raven, and R. W. Thorp. 1973. Comparative behavior of bees and Onagraceae. IV. *Clarkia* bees of the western United States. *Univ. Calif. Publ. Entomol.* 70:1–80.

Manning, A. 1956. Some aspects of the foraging behavior of bumblebees. *Behaviour* 9:164–201.

Menzel, R., J. Erber, and T. Masuhr. 1974. Learning and memory in the honeybee. In *Experimental analysis of insect behavior*, ed. L. Barton Browne. New York: Springer-Verlag.

Michener, C. D. 1974. *The social behavior of the bees.* Cambridge: Belknap Press of Harvard University Press.

Milliron, H. E. A monograph of the Western Hemisphere bumblebees. *Entomol. Soc. Can. Suppl.* 65 (literature review on bumblebees till 1961). *Entomol. Soc. Can. Mem.* 82:1–80 (1971); 89:81–237 (1972); 91:239–333 (1973).

Milliron, H. E., and D. R. Oliver. 1966. Bumblebees from Northern Ellesmere Island, with observations on usurpation by *Megabombus hyperboreus* (Schönh.) (Hymenoptera: Apidae). *Can. Entomol.* 98:207–213.

Mitchell, T. B. 1962. *Bees of the eastern United States*, vol. 2. North Carolina Agr. Exp. Sta. Tech. Bull. No. 152.

Morse, D. H. 1977. Resource partitioning in bumble bees: the role of behavioral factors. *Science* 197:678–679.

Müller, H. 1873. *Die Befruchtung der Blumen durch Insekten.* Leipzig: Engelmann.

Müller, H. 1881. *Alpenblumen.* Leipzig: Engelmann.

Oster, G., and B. Heinrich. 1976. Why do bumblebees "major"?: A mathematical model. *Ecol. Monogr.* 46:129–133.

Park, W. 1922. Time and labor factors involved in gathering pollen and nectar. *Am. Bee J.* 62:254–255.

Pendrel, B. A. 1977. The regulation of pollen collection and distribution in bumble bee colonies (*Bombus* Latr.: Hymenoptera). M.S. thesis, Department of Zoology, University of Toronto.

Plath, O. E. 1934. *Bumblebees and their ways.* New York: Macmillan.

Plowright, R. C., and S. C. Jay. 1966. Rearing bumble bees in captivity. *J. Apic. Res.* 5:155–165.

Plowright, R. C., and S. C. Jay. 1977. On the size determination of bumble bee castes (Hymenoptera: Apidae). *Can J. Zool.* 55: 1133–1138.

Pojar, J. 1974. Reproductive dynamics of four plant communities of southwestern British Columbia. *Can. J. Bot.* 52:1819–1834.

Proctor, M. C. F., and P. F. Yeo. 1972. *The pollination of flowers.* London: Collins.

Reader, R. J. 1975. Competitive relationships of some bog ericads for major insect pollinators. *Can. J. Bot.* 53:1300–1305.

Reader, R. J. 1977. Bog ericad flowers: self-compatibility and relative attractiveness to bees. *Can. J. Bot.* 55:2279–2287.

Ribbands, C. R. 1952. The relation between the foraging range of honeybees and their honey production. *Bee World* 34:2–6.

Richards, K. W. 1973. Biology of *Bombus polaris* Curtis and *B. hyperboreus* Schönherr at Lake Hazen, Northwest territories (Hymenoptera: Bombini). *Quaest. Entomol.* 9:115–157.

Richards, O. W. 1968. The subgenus divisions of the genus *Bombus*

Latreille (Hymenoptera: Apidae). *Bull. Brit. Mus. Nat. Hist.* (Entomol.) 22:211–276.

Röseler, P. F., and I. Röseler. 1974. Morphological and physiological differentiation of the castes in the bumblebee species *Bombus hypnorum* (L.) and *Bombus terrestris* (L.) *Zool. Jb. Physiol.* 78:175–198.

Sakagami, S. F. 1976. Specific differences in the bionomic characters of humblebees: A comparative review. *J. Faculty of Science, Hokkaido Univ. Zool.* 20:390–447.

Sakagami, S., and R. Zucchi. 1965. Winterverhalten einer neotropischen Hummel, *Bombus atratus,* innerhalb des Beobachtungskastens: Ein Beitrag zur Biologie der Hummeln. *J. Faculty Science, Hokkaido Univ. Zool.* (Ser.6) 15:412–762.

Schwintzer, C. R., and G. Williams. 1974. Vegetation changes in a small Michigan bog. *Am. Midl. Nat.* 92:447–459.

Sladen, F. W. L. 1912. *The humble-bee.* London:Macmillan.

Small, E. 1976. Insect pollinators of the Mer Bleue Peat Bog of Ottawa. *Can. Field-Natur.* 90:22–28.

Smith, Adam. 1776. *An inquiry into the nature and causes of wealth of nations.* London:W. Strahan and T. Cadell.

Stephen, W. P. 1957. *Bumble bees of western America (Hymenoptera:Apoidea).* Oregon State College Agr. Expt. Sta.: Tech. Bull. No. 40. Corvallis.

Stiles, F. G. 1975. Ecology, flowering phenology, and hummingbird pollination of some Costa Rican *Heliconia. Ecology* 56:285–301.

Szabo, T. I., and D. H. Pengelly. 1973. The over-wintering and emergence of *Bombus (Pyrobombus) impatiens* (Cresson) (Hymenoptera: Apidae) in Southern Ontario. *Insectes Soc.* 20:125–132.

Thien, L. B., and B. G. Marcks. 1972. The floral biology of *Arethusa bulbosa, Calopogon tuberosus,* and *Pogonia ophioglossoides* (Orchidaceae). *Can. J. Bot.* 50:2319–2325.

Thorp, R. W. 1978. Bumblebees of California. (In press.)

Thorp, R. W., D. L. Briggs, J. R. Estes, and E. H. Erickson. 1975. Nectar fluorescense under ultraviolet irradiation. *Science* 189:476–478.

Wellington, W. G. 1974. Bumblebee ocelli and navigation at dusk. *Science* 183:550–551.

Wille, A. 1958. A comparative study of the dorsal vessels of bees. *Ann. Entomol. Soc. Amer.* 51:538–546.

Williams, N. H., and C. H. Dodson. 1972. Selective attraction of male englossine bees to orchid floral fragrances and its importance in long distance pollen flow. *Evolution* 26:84–95.

Wilson, E. O. 1971. *The insect societies.* Cambridge: Belknap Press of Harvard University Press.

Witham, T. G. 1977. Coevolution of foraging in *Bombus* and nectar dispersing in *Chilopsis*: a last dreg theory. *Science* 197:593–595.

Zeuner, F. E. and F. J. Manning. 1976. A monograph on fossil bees (Hymenoptera: Apoidea). *Bull. Br. Mus. Nat. Hist.* 27(3).

Index

Bumblebees of North and Central America

The numbers are keyed to the plates (overleaf). The letters in brackets indicate subgenera and are keyed to the list "Genera, Subgenera, and Ranges of North American Bumblebees" in Appendix B, which also provides additional information on the identification of bumblebees.

Bombus of North America

1	*B. fraternus* [A]
2–3	*B. crotchii* [E]
4	*B. nevadensis auricomus* [C]
5	*B. nevadensis nevadensis* [C]
6	*B. morrisoni* [D]
7	*B. pennsylvanicus pennsylvanicus* [J]
8	*B. pennsylvanicus sonorus* [J]
9	*B. griseocollis* [D]
10	*B. perplexus* [G]
11	*B. bimaculatus* [G]
12	*B. impatiens* [G]
13	*B. affinis* [A]
14	*B. vagans vagans* [G]
15	*B. lucorum lucorum* [A]
16–17	*B. sandersoni* [G]
18	*B. frigidus* [G]
19	*B. ternarius* [G]
20	*B. terricola terricola* [A]
21	*B. caliginosus* [G]
22	*B. vandykei* [G]
23	*B. vosnesenskii* [G]
24	*B. californicus* [I]
25	*B. terricola occidentalis* [A]
26–28	*B. rufocinctus* [F]
29	*B. bifarius nearcticus* [G]
30	*B. bifarius bifarius* [G]
31	*B. balteatus* [H]
32	*B. hyperboreus* [H]
33	*B. polaris polaris* [H]
34	*B. edwardsii* [G]
35	*B. melanopygus* [G]
36	*B. borealis* [I]
37	*B. fervidus fervidus* [J]
38	*B. appositus* [I]
39	*B. sylvicola* [G]
40	*B. huntii* [G]
41–42	*B. flavifrons dimidiatus* [G]
43	*B. flavifrons flavifrons* [G]
44	*B. centralis* [G]
45	*B. sitkensis* [G]
46	*B. mixtus* [G]

Bombus of Mexico and Central America

47	*B. dahlbomii*
48	*B. volucella*
49	*B. brachycephalus*
50	*B. mexicanus*
51	*B. medius*
52	*B. pullatus*
53	*B. ephippiatus*

Psythirus of North America

54	*P. fernaldae*
55	*P. variabalis*
56	*P. insularis*
57	*P. crawfordii*
58	*P. suckleyi*
59	*P. ashtoni*
60	*P. citrinus*